THE HANDBOOK OF MEDICAL HALLUCINOGENS

PSILOCYBIN THERAPY, MICRODOSING LSD, AND OTHER PSYCHEDELICS

FUNCTIONAL HEALTH SERIES

SAM FURY

www.SFNonfictionBooks.com

WARNINGS AND DISCLAIMERS

CONTENTS

Get Your Next SF Nonfiction Book FREE!

Claim the book of your choice at:

www.SFNonfictionBooks.com

You will also be among the first to know of all the latest releases, discount offers, bonus content, and more.

Go to:

www.SFNonfictionBooks.com

Thanks again for your support.

INTRODUCTION

A sensation of well-being and renewed life flowed through me. Breakfast tasted delicious and gave me extraordinary pleasure. When I later walked out into the garden, in which the sun shone now after a spring rain, everything glistened and sparkled in a fresh light. The world was as if newly created. All my senses vibrated in a condition of highest sensitivity, which persisted for the entire day.

–Albert Hofmann

Overview of Medical Hallucinogens

Hallucinogens, also known as psychedelics, are a class of drugs that have the ability to profoundly alter the user's mood, cognition, and perception of reality. Examples of psychedelic substances, which can be natural or synthetic, include ketamine, peyote (or mescaline), phencyclidine (PCP), psilocybin, and lysergic acid diethylamide (LSD). Psychedelics and hallucinogens work on neurotransmitters in the brain to change one's sensory and physical experiences, often inducing hallucinations. They also create other psychological and physiological effects, depending on the chemical compounds of the substance.

Different psychedelics act on certain neurotransmitters in the brain to affect the user's perception of their surroundings. For example, substances like mescaline and LSD are serotonin agonists, which means they bind to and activate serotonin receptors in the brain, increasing the amount of serotonin released into the central nervous system (CNS). Serotonin is a neurotransmitter that affects one's sleep, hunger, mood, and sensory perception. Other substances, such as ketamine and PCP are N-methyl-D-aspartate (NMDA) glutamate antagonists, which means they block the activity of NMDA receptors in the brain. NMDA receptors are a sub-type of glutamate receptors. Glutamate is a neurotransmitter that controls emotions, perception of pain, environmental responses, and learning. By preventing glutamate from acting on the NMDA receptor,

NMDA antagonists reduce the excitatory processes in the brain, inducing various sedative effects.

Certain hallucinogenics have been used for millennia for cultural, spiritual, healing, and religious purposes, and as traditional medicine. Archeological evidence suggests that several Mesoamerican cultures, dating back to 5,000 years ago, used psychoactive substances in religious and healing ceremonies to induce altered states of consciousness, trances, and intoxication. Mushrooms containing psilocybin (*psilocybe mexicana*), and plants such as peyote (*lophophora williamsii*), which contains mescaline, and morning glory (*ipomoea corymbosa*), which contains lysergic acid, were used by Maya, Aztec, and Olmec tribes for ritual ceremonies (Carod-Atal, 2015). Many of these psychoactive substances are still used by traditional healers and shamans today. However, certain psychedelic and dissociative drugs are synthetically created to mimic the hallucinogenic effect of natural compounds.

As it stands, many hallucinogens are illegal around the world, even though they are used quite commonly as recreational drugs. Because of this, there are still certain taboos attributed to their use. In addition, medical professionals are sometimes wary of condoning their use, and many don't consider any dosage safe for use. However, the effect of hallucinogens on brain neurotransmitters has ignited an interest in their use for the treatment of several psychiatric disorders like depression, post-traumatic stress disorder (PTSD), and anxiety.

While all hallucinogens have the potential to produce mind-altering experiences and change the perception of reality, moods, and cognitive processes, they are generally classified into three main groups based on their effect on the brain:

- **Psychedelic substances**: These include compounds such as psilocybin, LSD, mescaline, and N,N-Dimethyltryptamine (DMT). Psychedelics affect the processing of serotonin in the brain and induce hallucinations and changes in the sense of self. They may induce feelings of insightfulness.

- **Dissociative substances**: These substances affect the processing of glutamate in the brain and cause the user to experience feelings of 'disconnect' from their body and the environment. They include PCP and ketamine.
- **Other**: Substances like 3,4-methylenedioxy-methamphetamine (MDMA), ibogaine, and salvia can cause both psychedelic or dissociative effects in users.

The Current Resurgence of Medical Hallucinogens in Interest and Research

The use of LSD as a treatment for psychiatric disorders was first suggested in the 1940s, when it was further researched and introduced to commercial medicine. By the 1950s, active compounds were being isolated from psychoactive substances to treat mental illnesses. LSD was used by various prominent psychologists and psychiatrists to treat disorders such as minor personality disorders and alcoholism. However, two decades later, legislation was implemented in the US that restricted the use of hallucinogens in medical treatments.

Since then, interest in the use of hallucinogens to treat psychiatric disorders has boomed. In 2017, a landmark ruling by the US Food and Drug Administration (FDA) designated the use of MDMA as "breakthrough therapy," followed by the same designation of psilocybin in 2019. This expedited the development and review of these psychoactive substances for medical use to treat serious conditions (*Psychedelic medicine*, 2023, para. 1).

Used at relatively low doses, studies have shown that patients generally don't experience major negative side effects when taking hallucinogens besides nausea, a slight increase in heart rate, and minor dizziness. Rather, patients tend to show the immediate diminishing of symptoms of anxiety and depression following the administration of hallucinogens. Certain psychedelic substances have also been identified for use in end-of-life care.

While initial studies focused on small sample sizes and a limited range of hallucinogenic substances, recent interest in their beneficial effects on psychiatric disorders has triggered an increase in clinical trials to treat depression, anxiety, and addiction. This is especially pertinent for patients who have not yet received relief from these psychiatric disorders through the use of conventional medication.

Experts suggest that micro-dosing of hallucinogens, which is administering a dose that is below the hallucinogenic threshold, has great potential to treat the symptoms of depression and anxiety without the negative side effects of conventional drug treatments.

Clinical trials on the use of psilocybin to treat depression have shown that a single dose of the hallucinogen produces marked improvement in symptoms of patients suffering from major depressive disorder. In addition, patients experience significantly fewer negative side effects from the treatment. Separate double-blind studies on the use of psilocybin to assist with tobacco addiction showed an 80% success rate, as opposed to behavioral and pharmacological therapies (less than 35% cessation of smoking) (*Psychedelic medicine*, 2023, para. 3).

The increase in mental health issues across the country has prompted many experts to identify this as a medical emergency. The use of hallucinogenic substances to treat complex mental health disorders is showing promise and has provided hope for those suffering from these difficult-to-treat conditions.

The goal now is to ensure regulated studies, rigorous scientific protocols, and consistent ethical practices to guide and monitor the safe use of hallucinogens to treat serious psychiatric disorders.

Trigger Warning

This book will deal with various substances that are commonly used as recreational drugs and may be illegal in the United States. It contains descriptions of the effects of these substances and examples and consequences of inappropriate use. It also describes the effects of drug abuse and addiction. The book references and describes various psychiatric disorders, such as PTSD, depression, and anxiety.

UNDERSTANDING HALLUCINOGENS

The History of Hallucinogens

In the past, the effects of plant-based hallucinogenics on the mind, as well as their ability to trigger various ecstatic and visionary states, led them to be described as having mystical or magical properties. Ancient Mesoamerican and North American cultures like the Aztecs, Native Americans, and Amazonian Indians used these hallucinogenic plant extracts solely in religious and spiritual rituals, due to their divine associations.

Psilocybin and LSD

The Aztecs used teonanácatl (*psilocybe mexicana*), a sacred mushroom also called Flesh of the Gods, to achieve the effects of psilocybin (4-phosphoryloxy-N,N-dimethyltryptamine) in healing and religious rituals. Teonanácatl was used to induce visions and a state of trance to allow shamans and healers to communicate with the gods. While stories of the use of psilocybin are largely ascribed to oral lore, there is evidence of the use of psychotropic substances by the Aztecs as recorded by Spanish missionaries in the 16th Century.

In 1943, Swiss chemist Albert Hoffman, discovered LSD, a hallucinogen that is a semisynthetic derivative of the chemical compound ergotamine. Ergotamine is a form of naturally occurring alkaloid found in ergot fungi. Psilocybin was only introduced to the Western world in the 1950s, largely facilitated by American ethnomycologist Robert Gordon Wasson. His article in a 1957 edition of LIFE magazine sparked interest in the psychedelic effects of plants. Later, Timothy Leary, an American psychologist and author, started his research and experimentation on the effects of these psychedelic substances, most notably LSD. After the 1950s, research on the effects of both natural and synthetic hallucinogens surged, only halted by a ban on research on psilocybin in the 1970s. However, this ban was recently lifted, and psilocybin is now becoming touted as a potential treatment for a variety of addictions and dependencies, as well as other psychiatric disorders.

Mescaline

Mesoamerican Indians used the peyote cactus to induce supernatural visions in religious and healing ceremonies. Records show its use by indigenous cultures dating back to 1000 BC. It was used again in North America in the late 1880s, whereafter it was banned in 1880 by various US states. In 1897, the compound responsible for the hallucinogenic effects of peyote, namely mescaline, was isolated from the cactus by Arthur Heffter, a German chemist. From there it was synthesized in the lab in 1919.

The use and possession of peyote was banned in several states in the US in 1930. However, the 1950s saw an increase in its use in psychiatric treatments like schizophrenia. In 1954, the English writer and philosopher, Aldous Huxley, coined the term "psychedelic" in response to a hallucinogenic experience on mescaline. Yet in 1970, a bill was passed that labeled mescaline an illegal Schedule 1 substance with no medical benefits. Despite this censure, mescaline used in religious ceremonies is now exempt from this classification under the 1994 American Indian Religious Freedom Act (AIRFA). Mescaline is also found in the San Pedro cacti (*trichocereus macrogonus var. pachanoi*) from Peru and Ecuador. Scientists are now able to synthetically create the chemical compound found in mescaline.

DMT

For thousands of years, Amazonian Indians in South America used plants like chacruna (*psychotria viridis*) for their hallucinogenic effects in religious ceremonies and rituals. The psychedelic effects of these plants are a result of DMT, a white powder isolated from chacruna and other plants, like the yopo or calcium tree (*anadenanthera peregrina*). DMT is consumed in various forms, from smoking to drinking in a brew called ayahuasca. It is still used today in various religious ceremonies to obtain spiritual awakening and insight. It is illegal to manufacture, possess, buy, or distribute DMT in the US, as it is a Schedule 1 controlled substance. However, if approved by the Drug

Enforcement Administration (DEA) and the Food and Drug Administration (FDA), researchers can use DMT under the Schedule 1 research registration.

The Science Behind Hallucinogens

There are slight differences in the way that hallucinogens work on the brain. As discussed in the Introduction, there are three main categories of hallucinogenic substances, namely psychedelics, dissociatives, and other drugs. The affected area of the brain will influence how the substance is used to treat various psychiatric conditions.

1. ***Psychedelic drugs:*** The chemical composition of classic psychedelic drugs like LSD and psilocybin are similar to that of the neurotransmitter serotonin. Because of this, they are classed as serotonin agonists, which means they activate serotonin receptors, more specifically the 5-HT receptor, which is the most common type. The binding of psychedelic compounds (especially LSD) to these receptors causes the release of neurotransmitters like serotonin, dopamine, and glutamate. The significant increase in these neurotransmitters triggers activity in the prefrontal cortex of the brain, specifically in the right brain hemisphere, which controls spatial perceptions, emotions, and nonverbal communication like body language and facial expressions. This is why psychedelic substances may cause the user to experience dilated pupils, increased heart rate, raised blood pressure, and muscle tremors.

   The altered sense of self that is often an effect of psychedelics is a result of the disruption of the Default Mode Network (DMN). The DMN is a set of brain areas that work together when the mind is at rest, or in a passive state. This includes states of daydreaming, mind-wandering, and introspection. Psychedelics reduce the interconnections between the regions of the brain that form the DMN, causing the user to experience episodes of ego dissolution and a loss of separation between the self (a decrease in the identification of self) and others, including the surrounding environment. The combination of these effects is why psychedelic (mind-manifesting) substances are also called psycholytics (psyche-loosening). Because of this, current research on psychedelics is promising for their use in treating depression.

2. ***Dissociative drugs***: Substances like PCP and ketamine are common hallucinogens with dissociative properties. They work primarily on the NMDA receptors of the brain (a subtype of glutamate receptors) and are glutamate antagonists, which means they inhibit the function of glutamate. By blocking NMDA receptors, normal neural communication is disrupted, affecting cognitive functions like learning, memory, emotion, mood regulation, and pain perception. Research suggests that overexposure of the brain to high levels of glutamate over time can lead to depression. So, using dissociative drugs like ketamine can decrease the production of glutamate, producing an anti-depressive response in users. In low doses, dissociative substances may produce feelings of confusion, and disorientation, as well as sensory distortions, hallucinations, and feeling detached from self. At higher doses, the user may experience loss of memory, increased psychological distress, such as panic, anxiety, aggression, and paranoia, and extreme physical reactions like harmful changes in heart rate, blood pressure, and body temperature.

Because glutamate is responsible for interpreting pain, dissociative substances that inhibit glutamate production are used to reduce pain in the form of anesthetics and analgesics. Dissociative substances can also increase the release of dopamine in the brain, leading to feelings of euphoria and pleasure. They are under review for their use in treating clinical depression and suicidal ideation. Ketamine is also being highlighted as a potential treatment for obsessive-compulsive disorder (OCD).

3. ***Other forms of hallucinogenic drugs***: Substances such as MDMA and ibogaine work on different parts of the brain, sometimes inducing similar effects to those of both dissociative and psychedelic substances. While mostly classified as "other hallucino-

genic drugs," they are chemically known as entactogens or empthogens.

Entactogens, like MDMA (also known colloquially as ecstasy), appear to cause deactivation of the amygdala and thalamus, combined with the activation of the regions of the brain that control cognitive, emotional, and behavioral functions. These structures include the prefrontal cortex, limbic system, and paralimbic structures. Stimulation of this area of the brain results in the heightened positive mood associated with entactogenic substances. Entactogens also increase the user's feelings of empathy towards others, as well as friendliness and playfulness. Because entactogens decrease the levels of serotonin in the brain, they can also cause depression, mood swings, and anxiety. Generally, however, entactogens are associated with increased energy, warmth, sexual arousal, and a sense of belonging and connectedness.

MDMA, which is a synthetic substance, can induce stimulant effects, and may also alter perceptions of time, a psychedelic characteristic. Ibogaine also has various psychedelic properties but affects different areas of the brain. MDMA is being touted as a treatment for post-traumatic stress disorder.

The effects of different types of hallucinogens are based on their stimulation of serotonin receptors, specifically the serotonin 2A receptor. Studies suggest that the serotonin 2A receptor is the primary receptor that controls the effects of hallucinogens on the brain. This is called the affinity of hallucinogens to the serotonin 2A receptor. For example, hallucinogens with a strong attraction to this receptor, like LSD, show a significantly higher potency than others. Hallucinogenic substances act to mimic the effect of serotonin on these receptors by binding to them and altering their conformation. This then changes the behavior of the neuron on which that receptor sits, the excitatory pyramidal neuron.

Pyramidal neurons are found in the mammalian cerebral cortex of the brain, as well as in the hippocampus and amygdala. They are some of the most abundant neurons in the brain, comprising up to two-thirds of the cerebral cortex. The main role of pyramidal neurons is to translate the inputs received at the synapses into output as action potentials, in other words, to facilitate the transmission of information. However, they are slightly different from other neurons in their long axons, some of which actually traverse areas of the nervous system out of the brain.

There are two main types of pyramidal neurons, namely excitatory and inhibitory neurons. Excitatory neurons release glutamate, a neurotransmitter that promotes action potentials in neurons, allowing information to be passed between nerve cells. Because of this, it is essential in learning, behavior, memory, and other cognitive processes, as well as for the ability of synapses (the junctions between nerves) to change (synaptic plasticity). On the other hand, inhibitory pyramidal neurons release gamma-aminobutyric acid (GABA), which is a neurotransmitter that inhibits the transmission of chemical messages amongst neurons in the nervous system.

Type 2A serotonin receptors are primarily found on excitatory pyramidal neurons, specifically those found in the region of the cerebral cortex closest to its base, which is called layer five. These excitatory neurons connect liberally with inhibitory interneurons within that area. Because inhibitory neurons release GABA, which inhibits the action of other neurons, the net effect of the release of glutamate from excitatory neurons is thought to be inhibitory to other neurons in the cerebral cortex. Simply put, neuroscientists now believe that, overall, the stimulation of excitatory neurons actually has a dampening effect on the transmission of information in the cortex.

This discovery needs to be taken into account when studying the effects of hallucinogens on the brain. While activation of the 2A serotonin receptors by these substances was previously believed to have an excitatory effect, research now suggests that the effect of hallucinogens on layer five pyramidal neurons is inhibitory. In other words, hallucinogens cause a decrease in brain activity, specifically in the thalamus, medial prefrontal cortex, and posterior cingulate cortex. This causes the changes in consciousness that are commonly seen in users of hallucinogens. While this doesn't mean the brain works less under the influence of hallucinogens, it does suggest that there is greater flexibility and interconnectedness within the brain when exposed to these substances.

Basically, the excitatory action of hallucinogenics causes a disorganizing influence on the activity of the cortex as a whole. Important neurons in the cortex are prompted to fire out of sync with the larger population of neurons, which usually work together in a rhythmic pattern. This is because hallucinogens cause networks within the brain to communicate more openly with one another, in a less restricted manner, promoting enhanced cognitive flexibility. It is this increased cognitive flexibility that explains the effectiveness of hallucinogens in supporting cognitive-based psychotherapies in the treatment of disorders such as OCD, addiction, and depression. The flexibility of cognitive processes within the brain allows for entrenched thoughts and behaviors, accredited to various psychiatric disorders, to be altered.

One of the most fascinating aspects of hallucinogens and their impact on the mind is the dissolution of self, or what is colloquially known as "ego death." This is found by users of psychedelics such as psilocybin, who report experiencing the disintegration of ego, or a feeling of loss of self-identity and complete union with the world. Studies on this phenomenon show a significant correlation between reports of ego disintegration and decreases in electrical activity in the posterior cingulate cortex (PCC). The posterior cingulate cortex is the region of the brain involved in introspection, daydreaming, and the relating of information from the external world to the self. It appears as though hallucinogenics disrupt the patterns of activity in the PCC to prompt ego-disintegration.

While hallucinogenic substances have historically been labeled dangerous and medically irrelevant, current research is starting to highlight their usefulness in treating various mental health disorders and illnesses. Researchers and scientists are working to decode the chemical properties of these substances and how this determines their therapeutic effects. The goal is to better understand the mechanisms behind these substances in order to develop treatments that are safe and effective.

Neuroscientists note, however, that the use of hallucinogens needs to be done within a holistic therapeutic approach. Boris Heifets, an anesthesiologist from the Wu Tsai Neurosciences Institute, suggests that hallucinogens are a catalyst for recovering from various disorders, rather than actual treatments (Kim, 2023). In fact, many studies on hallucinogens suggest that, because these substances promote brain-altering experiences, their effects are more deeply connected to the therapeutic results of these introspective trips, realignments of values, or changes in perspective (Kim, 2023). For example, patients with PTSD that are treated with MDMA show a decrease in fear responses. It is believed that this reduction in fear is a result of the mind-altering effects of MDMA that allow patients to shift their perspectives on their trauma in a safe and grounded state. However, researchers are still working to isolate the effects of hallucinogens so that their benefits can be achieved without their psychedelic associations.

Unfortunately, the potential for misuse of hallucinogenic substances means that widespread medical legalization may create liability issues. However, medical supporters of these treatments suggest they may provide relief to those suffering from mental health disorders that have so far resisted conventional medicine.

The Legal Battle Over Psychedelics

Despite their historical and traditional uses, as well as their popularity as both recreational drugs and treatments for various psychiatric disorders, psychedelics were outlawed all over the US by the year 1970. Concerned that they were becoming a menace to society, most of them were categorized as Schedule 1 substances under the new federal Controlled Substances Act (CSA). The category of Schedule 1 substances defines compounds that, among other aspects, have no accepted medical use. Later, psychedelics were banned for recreational use, globally, via the United Nations Convention on Psychotropic Drugs.

Sproul (2021) reviewed the history behind the criminalization of psychedelic drugs, specifically in the US, summarizing it as follows:

Timeline	Historical Events
≥ 3,000 years ago	Naturally occurring psychedelics such as psilocybin, DMT, ibogaine, and mescaline were used by Indigenous cultures to enhance and assist in religious and spiritual ceremonies and rituals. These compounds were not used recreationally; rather they were revered as sacred conduits to the divine.
Late 19th Century	White Americans start experimenting with peyote, the plant that contains the psychoactive compound, mescaline.
1897	German chemist and pharmacologist, Arthur Heffter, isolates mescaline from the peyote cactus.
1912	MDMA (ecstasy) is synthesized by Merck, a multinational pharmaceutical company.
1938	Swiss chemist, Albert Hofmann, synthesizes lysergic acid diethylamide or LSD.
1943	LSD's psychoactive properties are discovered by Hofmann.
1947	LSD is marketed by pharmaceutical company, Sandoz, as Delysid, for treatment of various psychiatric disorders. It is classified as a psychotomimetic for its believed ability to induce a temporary state of psychosis.
1940s and 1950s	The CIA implemented MK-ULTRA, a top-secret research project that tested LCD in an attempt to discover a truth serum, and as a tool of brainwashing. PCP is marketed as an anesthetic but is later taken off the market as a result of its hallucinogenic side effects. Psychedelics such as LSD, mescaline, and psilocybin are researched for their use in psychotherapy.
1957	Psychiatrist, Dr. Humphrey Osmond and his colleague, Dr. Al Hubbard, establish LSD treatment centers throughout the US to treat alcoholism.
1959	Public interest in psychedelics grow, with extensive use by all walks of life across the US, including celebrity endorsements and in psychedelic therapy. LSD is regarded as a treatment for alcohol addiction.
1960	Timothy Leary, a Harvard psychology professor, uses psilocybin on a visit to Mexico, returning to start the Harvard Psilocybin Project.
1962	The FDA halts the Harvard Psilocybin Project, as well as imposing strict new regulations on the use of psychedelics in pharmaceutical research. This leads to a sharp decline in the research into psychedelics.
1965	The FDA bans the sale and unlicensed manufacturing of LSD through the Drug Abuse and Control Amendments. However, the use of psychedelics soars with the Free Speech Movement and hippie counterculture.
1966	Sandoz cuts off the last remaining source of legal LSD by ending both the production and sale of the drug. Various supporters of psychedelics in therapeutic applications oppose their restriction, including Senator Robert Kennedy, Timothy Leary, Allen Ginsberg, and Dr. Stanley Yolles.
1968	The US Federal Government bans the possession of LSD.

1970	LSD and other psychedelic substances are categorized as Schedule 1 drugs, under the new Controlled Substances Act. Schedule 1 drugs have no medical applications and have the highest potential for abuse.
1971	The UN Convention on Psychotropic Substances places psychedelics in the highest category, alongside heroin, and above cocaine.
1980s	Despite their criminalization, psychedelics are widely used in recreational settings. MDMA is popular in nightclubs.
1985	MDMA is categorized as a Schedule 1 substance.
1970-2000s	Most research applications are denied by the FDA. The blanket ban on psychedelics all but halts research into their therapeutic uses.
1996	The Beckley Foundation is instituted, which researches LSD, psilocybin, MDMA, and DMT.
2000s	The Multidisciplinary Association for Psychedelic Studies (MAPS), which was founded in 1986, advocates for research into the therapeutic use of psychedelics to treat psychiatric disorders.
2006	The Johns Hopkins University publishes a milestone study that revives research into psilocybin.
2013	MDMA-assisted psychotherapy for PTSD is designated as a Breakthrough Therapy by the FDA.
2016	The first-ever brain imagery study is done, analyzing the brain on LSD.
2018	Johns Hopkins University researchers promote the re-categorization of psilocybin from a Schedule 1 to a Schedule 4 substance. Psilocybin-assisted psychotherapy for treatment-resistant depression is designated as a Breakthrough Therapy by the FDA.
2019	Denver, Colorado, becomes the first city to decriminalize psilocybin. The Johns Hopkins Center for Psychedelic and Consciousness Research is launched.
2020	Oregon becomes the first state to decriminalize and legalize psilocybin for therapeutic use.
2021	More than $730 million is invested by private businesses in the research and development of psychedelic drugs in therapeutic treatments. Johns Hopkins Medicine is awarded a grant to research the use of psilocybin to treat tobacco addiction, the first of its kind in 50 years.
2022-current	Several ballot initiatives and legislative efforts are promoted across the US, with the intention of legalizing and decriminalizing psychedelics for medical and therapeutic use.

PSILOCYBIN

As a naturally occurring alkaloid found in many mushroom species, psilocybin induces a hallucinogenic effect in users. Fungi containing this chemical are often colloquially called "magic mushrooms." Psilocybin mushrooms are recreationally used as a psychedelic substance to induce hallucinations and states of euphoria. The DEA lists it as a Schedule 1 substance, which are chemicals that currently have no accepted medical use and have a high potential for abuse. This includes other drugs such as peyote, LSD, and cannabis. Although there are no substantiated records of addiction to psilocybin, users may experience unpleasant reactions to the substance like panic attacks and anxiety. This is one of the reasons it is still considered an illegal drug. Nonetheless, research is growing on the potential effects of psilocybin on treatment-resistant major depressive disorder.

In a ground-breaking case in 2021, Johns Hopkins Medicine was awarded the first-ever federal grant in over fifty years from the National Institutes of Health (NHI) to study the effects of psilocybin therapy on tobacco addiction. Johns Hopkins Medicine works closely with the University of Alabama and New York University to study how psychedelics affect mood, brain function, cognitive processes, and behavior. The research aims to understand the use of psilocybin and other psychedelics to treat illness and disorders including PTSD, Alzheimer's disease, anorexia nervosa, addiction, major depression, and opioid addiction, amongst others (*Psychedelics research*, n.d., para. 15).

Chemistry and Pharmacology

The most popular or well-known source of psilocybin is the teonanácatl mushroom (*psilocybe mexicana*) from Mexico, an ancient and sacred fungi used in cultural and spiritual rituals. The compound responsible for the hallucinogenic effects of this mushroom is also found in other species, including *psilocybe semilanceata* in the UK, and *stropharia cubensis* and *conocybe* spp. in the US. Mushrooms containing psilocybin will often bruise blue when handled. Over 15 varieties of mushrooms contain the chemical; however, some of them are poisonous, so care needs to be taken when ingesting them.

The compound psilocybin was first described in Western medicine by American biologist and ethnobotanist, Richard Evans Schultes. In 1938, Schultes visited Oaxaca, a city in southern Mexico, after hearing about a psychedelic compound occurring naturally in fungi that was used by the Mazatec people to induce divination and hallucinations. In 1955, American ethnomycologist Robert Gordon Wasson and his wife Valentina Wasson, also an ethnomycologist, traveled to Oaxaca where they participated in an ayahuasca ceremony. Wasson later recorded his experience in LIFE magazine, which spurred great interest amongst the American public.

The chemical structure of psilocybin is 4-phosphoryloxy-N,N-dimethyltryptamine. The metabolized compound of that, psilocin, has a chemical structure of 4-hydroxy-N,N-dimethyltryptamine. Psilocybin needs to be metabolized into psilocin in order for it to be absorbed into the brain. It is psilocin that is so similar in chemical structure to serotonin. Because psilocin is highly lipophilic, which means it is soluble in lipids, fats, and oils, it can easily cross the blood-brain barrier and bind effectively with 5-HT 2A receptors in the brain. This is in comparison with psilocybin, which is hydrophilic. If taken orally, psilocybin is broken down by stomach acids to form psilocin. Psilocybin can also be metabolized in the intestines, blood, and kidneys to produce psilocin.

Both psilocybin and psilocin are non-specific partial agonists, which means they are not highly selective to the type of receptor that they can bind to. It also means that these compounds have a broader range of effects, making it difficult to accurately predict the effects of these substances on the brain. Because of this, and the lipophilic nature of psilocin, well-managed clinical therapies should advise patients to take the substance on an empty stomach to better control the outcomes.

The compound psilocybin shares a chemical structure with other hallucinogens like DMT and LSD, as well as with neurotransmitters like serotonin and melatonin. This chemical structure is called tryptamine, which is a metabolic derivative of the amino acid tryptophan. As such, hallucinogens like psilocybin are often referred to collectively as tryptamines. Tryptamines are alkaloids naturally occurring in plants, fungi, and animals, as well as in the human brain. These organic compounds are composed of an indole, which is a fused double ring of benzene and pyrrole, with an added side chain of 2-carbon. Tryptamines are 5-HT 2A receptor agonists, which will be discussed later in this chapter. They cause various sensory hallucinogenic effects by interacting with serotonin in the brain. Synthetic modifications of the tryptamine chemical structure can either reduce or increase the potency of the compound.

With the agonistic effect of psilocin on 5HT-2A receptors in the cortex, there is a responsive release of glutamate. This is significant, specifically in patients with alcohol addiction and other dependencies. The pathway along which information passes from the prefrontal cortex (PFC) to the nucleus accumbens (a region of the brain that manages addiction and reward) is abundant with these receptors. Those with an addiction to alcohol show a deficiency in that pathway. In a small study on the effects of psilocybin on alcohol addiction, participants showed a significant decrease in alcohol cravings from the baseline level established pre-treatment. In addition, there was a 27% decrease in the number of days of alcohol consumption, in comparison to that pre-treatment (Ziff et al., 2022).

The chemical compound in psilocybin activates centers in the brain that manage mood, perception, and cognition, as well as centers of arousal and panic (Davis, 2023). These include the PFC, the thalamus, and the limbic centers. The PFC works to integrate information received from the amygdala and hippocampus. This integration allows the brain to produce thoughts of self- reflection and also assists in other cognitive processes, such as decision-making and planning.

When a person is suffering from depression, the amygdala and PFC become hyperactive, forming excessive negative ruminations that muddle memories associated with introspection. Psilocybin acts on the thalamus by altering signals to the brain that process sensory information. This is why users experience various sensory hallucinations and changes in perception and integration of sensory information. The effect of psilocybin on the amygdala and hippocampus, which form the limbic system, causes changes in stress responses, the formation and retrieval of memory, and the regulation of emotion. It is the ability of psilocybin to encourage the plasticity of neural networks that may be the underlying factor in its effectiveness to ease depression.

Psilocybin mushrooms are generally taken orally, at which point this chemical compound is metabolized to psilocin, which is the primary active chemical (*psilocybine*, 2019, para. 6). Psilocybin is metabolized into psilocin in the intestines. It is psilocin that has the psychoactive properties. In general, dried mushrooms contain about 0.5% psilocybin. The alkaloids responsible for the hallucinogenic effect of psilocybin are found predominantly in the cap of the mushroom, rather than the stem. The onset of the hallucinogenic effect of the chemical is within twenty minutes to one hour, after which the effects decrease from three to six hours. Clinical studies show that orally ingested psilocybin shows up in the bloodstream as psilocin approximately thirty minutes after administration. General positive effects of the chemical are feelings of energy, insight or spiritual awareness, increased emotional sensitivity, and hallucinogens. Negatively perceived experiences include blackouts, confusion, paranoia, anxiety, fear, and nausea.

Therapeutic Applications of Psilocybin

Small clinical studies done on the effect of psilocybin on major depressive disorders have shown that just a single dose of the substance can significantly decrease the symptoms of depression in people suffering from medically resistant disorders. These findings have encouraged the US Drug Enforcement Administration (DEA) to label psilocybin as a breakthrough medicine (LaMotte, 2023). To date, psilocybin has been used in the treatment of a variety of psychiatric disorders such as anxiety, obsessive-compulsive disorder (OCD), anorexia, depression, and addiction.

Because psilocybin mimics the chemical structure of serotonin, it has similar effects on the brain to that neurotransmitter. Other studies have shown strong evidence for the positive effects of psilocybin on cancer-related depression, as well as treatment-resistant depression. This is based on multiple clinical randomized controlled trials. Some evidence is available for the treatment of addictions such as alcohol dependence and tobacco addiction, based on single-group trials, or open-label trials (where patients are aware of what they are taking, which can introduce treatment bias). However, little clinically significant evidence is associated with beneficial effects on the treatment of OCD and cluster headaches (McCallum et al., 2022).

To better understand the mechanism by which the psilocybin compound binds to serotonin receptors in the brain, scientists have used brain mapping techniques to produce high-resolution images of the chemical once it has actively attached to the relevant receptor. It is this research into the molecular activity of the compound that is assisting scientists in understanding how to increase the therapeutic benefits of psilocybin while decreasing the psychedelic effects.

The use of psilocybin to treat psychiatric disorders should be done under clinical, controlled conditions with a medical professional. This means that the potential for a negative experience should be reduced. However, due to the complex and controversial history of psychedelics like psilocybin, there is no clear-cut guidance for healthcare professionals on how to use them to treat psychiatric disorders. Initially, clinical trials dosed participants based on body weight, with three to four applications occurring every three to four weeks. However, new research suggests that psilocybin should be administered at a fixed dose, regardless of body weight. Studies show that there are no significant differences in the psychedelic effects between fixed-dose and weight-based applications. The current standard dose for therapeutic use is a fixed 25 mg of extracted isolate or synthetic psilocybin. If psilocybin-containing mushrooms are used, the standard dosage is 2.5 grams of whole or powdered dry *Psilocybe cubensis* mushroom. However, it must be kept in mind that the content of psilocybin in different mushroom species can vary from 0.5 to 2%, so it's best to start new patients with a lower dose than the standard dosage prescribed. Because psilocybin is metabolized in the stomach, liver, and intestines, sensitivities in the functions of these organs may alter the effect of the compound.

Additionally, any psilocybin-based treatment for psychiatric disorders should, in good practice, be integrated with assisted psychotherapy. This allows the participant to develop a goal and objective for the therapy, in other words, participate in an intention-setting activity. This is usually done in the pre-treatment session and provides the participant with an idea of what to expect during the session, as well as expected outcomes. During the treatment, sessions are often accompanied by tools such as meditation, breathwork, and music. Following treatment, the participant attends post-treatment sessions to process and integrate the experience.

There are several benefits of psilocybin as a treatment for psychiatric disorders (Ziff et al., 2002):

- Psilocybin has a proven low-abuse potential.
- There are no documented incidents of physical dependence.
- There are no documented incidents of withdrawal.
- The half-life of psilocin is three hours, which is more rapid than other psychedelic substances with longer durations of activity.
- Any side effects associated with psilocybin are generally mild and rarely prolonged.
- Only two or three doses are enough to have long-term or even permanent benefits.

Clinical Studies and Research

The interest in the use of psilocybin as a treatment for major depressive disorders that are resistant to conventional medication is growing in interest. It is only recently that major pharmaceutical companies have started looking at conducting clinical trials into psilocybin as an anti-depressant. Although still in its early stages, the research on psilocybin as a treatment for major depressive disorder and other psychiatric disorders is growing. This research aims to differentiate the doses of psilocybin required to have the necessary anti-depressant effect without the hallucinogenic and sympathomimetic effects (such as increases in blood pressure, heart rate, and tremors) of the substance at a high dose. Clarke (2007) notes that high doses of psilocybin, above 5.0 g can cause open eye hallucinations, while lower doses between 0.5 and 2.5 g can cause closed eye hallucinations. Recreational doses of psilocybin sit between 1-5 g of the dried mushroom. Dosages above 5 g are recorded as having numerous unpleasant side effects.

Brain imaging techniques, such as positron emission tomography (PET), which looks at blood flow and metabolic processes in the brain to map physiological activities, have been used to study the effect of psilocybin on neurotransmitter receptors. As previously noted, psilocybin is a 5-HT 2A receptor agonist which means it binds to and activates a type of serotonin receptor (5-HT receptors) in the brain, causing various physiological and psychological effects. In other words, psilocybin enters the brain via the same receptors as neurotransmitters like serotonin and melatonin.

A study by Mann (2023) showed that higher doses of psilocybin correlated with a higher frequency of receptor occupancy, with a dose over 14 mg causing an occupancy rate of over 60%. Conventional antidepressants like selective serotonin reuptake inhibitors (SSRIs) require an uptake value of over 70% to have any therapeutic effects on the patient. Because psilocybin and other psychedelics are receptor agonists, while SSRIs use a blockade action to increase serotonin levels, a lower dose of psilocybin would be required to have a notable therapeutic effect.

In 2022, in a double-blind trial, researchers treated a sample of 233 participants with a single dose of psilocybin, at different dose strengths (1 mg; 10 mg; and 25 mg) (Goodwin et al., 2022). Participants were requested to refrain from using additional psychotropic medication while participating in the trial. The psilocybin dosage was accompanied by psychotherapy over a period of three weeks. The study used the Montgomery-Åsberg Depression Rating Scale (a clinical measure of the severity of depression) to determine potential changes to the rating in patients with treatment-resistant depression. Results showed a significant decrease in depression scores in patients treated with the 25 mg dose in comparison to the 1 mg dose. However, there were several slightly concerning side effects experienced by 77% of the participants, including headaches, dizziness, and nausea. Nonetheless, this study suggests that psilocybin is a rapidly acting substance that is effective in treating medication-resistant depression.

Ziff et al. (2022) reviewed several studies that investigated the use of psilocybin for treating mood and substance use disorders. The focus was specifically on tryptamine, the naturally-occurring alkaloid that causes hallucinations and is found in psilocybin. A small open-study conducted by Johns Hopkins University showed that of the 15 participants who undertook integrated psilocybin therapy, 80% had remained abstinent from smoking after six months following the treatment (Ziff et al., 2022). Currently, there are ongoing studies to evaluate the effectiveness of the therapy on participants with cocaine, methamphetamine, and opioid addiction.

Practical Considerations in Therapy

The chemical compound of psilocybin works in a similar manner to serotonin, which is the brain's feel-good chemical. Serotonin elicits feelings of happiness and optimism. However, people who suffer from psychiatric disorders like depression, anxiety, addictions, and

PTSD have diminished levels of serotonin. Traditionally, these disorders have been treated with SSRIs, which act to increase the concentration of serotonin in the brain. Unfortunately, SSRIs can take weeks to work and sometimes have no effect at all. Researchers are positive about the effect of hallucinogens like psilocybin, especially because they appear to effect changes in the brain's neural circuitry within 30 minutes of administration.

While psilocybin can increase neural connections between certain regions of the brain, it disrupts other networks, creating what neurologists refer to as a disorganized brain. The normal pathways and boundaries that exist between visual, auditory, executive, and cognitive processes are broken down, fragmenting well-worn neural pathways. While people suffering from depression and anxiety constantly engage these neural circuits to cement negative thought patterns and anxious thoughts, hallucinogens like psilocybin are able to disrupt these patterns, allowing an alternative perception or viewpoint and encouraging a way out of negative beliefs.

In addition to the ability of hallucinogens to break down old, negative thought processes, research has shown that certain psychedelic substances like psilocybin actually encourage the growth of new connections within the brain. Called neuroplasticity, psychedelics can reinforce and solidify new and positive neural circuits.

One of the positive aspects of psilocybin as a treatment for major depressive disorder is that they don't seem to have the reinforcing properties of traditionally addictive substances. It also doesn't cause craving, withdrawal, or dependence. Nonetheless, researchers are concerned that the ability of psilocybin to induce mystical experiences could create problematic use or psychological dependence. At this stage in the research, it appears as though the effects of psilocybin are temporary, with symptoms of psychiatric disorders reappearing several months after treatment. It is these kinds of questions that scientists are investigating further.

One of the potential considerations to counter the temporary effect of hallucinogenics is to implement microdosing, whereby very small doses of psilocybin (between 0.1 and 0.3 grams) are delivered to the participant several times a week. Unfortunately, peer-reviewed scientific studies have yet to present positive findings in this regard. Ultimately, there is hope that psilocybin can be used to reduce levels of anxiety and improve the user's feelings of well-being, which can then lead to a brighter perspective on life (Smausz, et al. 2022).

While the benefits of psilocybin therapy appear overwhelmingly positive, researchers caution over- and misuse of the substance. It is important to administer the substance in conjunction with regular psychotherapy, as well as in a controlled environment with an experienced professional. Experts suggest that psilocybin treatment may not be suitable for those diagnosed with bipolar disorder, schizophrenia, borderline personality disorder, or other psychoses, as psychedelic experiences may expose or exacerbate those conditions (LaMotte, 2023).

In addition, psilocybin should not be administered in conjunction with SSRIs. Because psilocybin works similarly to SSRIs, in that it interacts with the serotonin system, there is a risk of what is known as pharmacodynamic drug interactions (McCallum et al., 2022). This means that certain drugs can interact with one another to alter the effects of the psychedelic. Depending on the psychotropic medication, whether it is an antipsychotic or antidepressant, the effect of psilocybin may be intensified or decreased.

Additionally, there is a risk of developing a life-threatening condition called serotonin syndrome when using psychotropic medication with psilocybin. Serotonin syndrome refers to a condition whereby a build-up of serotonin in the synapses of the brain can cause various symptoms from shivering and diarrhea to more severe effects like fever, seizures, muscle rigidity, and unconsciousness. Clinical trials will often require participants on SSRIs to taper and stop the use of these medications before starting with psilocybin therapy.

There are several adverse psilocybin-associated effects that should be considered when implementing this therapy (McCallum et al., 2022):

- **Common physical events**: increased blood pressure and heart rate, headache, nausea
- **Common psychological events**: anxiety and confusion
- **Less common physical events**: fatigue, migraine, vomiting, and physical discomfort
- **Less common psychological events**: extreme fear, paranoia, psychological discomfort, and psychotic-like symptoms

Further research is required to better understand the effects of psilocybin on the brain, using higher spatiotemporal resolution brain mapping techniques to measure the rate and frequency of activity generated by the action of psilocybin.

LSD

It was 1395, and a disease with terrifying symptoms was swarming through small European villages. Sufferers would find themselves dancing uncontrollably for days, or experiencing convulsions that lasted for hours. Symptoms further included vomiting, hallucinations, and pain that engulfed the extremities of the victims. These epidemics would decimate entire communities, sending fear and panic spreading like wildfire. People called the illness St. Anthony's fire, named after the monks who attempted to treat the victims. At the time, communities had no idea what caused the plague, sometimes blaming it on witchcraft.

In fact, certain historians believe that a similar illness was what caused the Salem witch trials. This was a period of time in 1692 when young girls were accused of bewitchment, after exhibiting symptoms that included convulsions, hallucinations, and temporary blindness. As a result, nineteen people, both men and women, were executed on suspicion of witchcraft. Researchers now suggest that the Salem witches, St. Anthony's fire, and other similar illnesses of the time were caused, in part, by a fungus called ergot. Ergot contains various alkaloidal structural groups, one of which is the lysergic acid amides. Lysergic acid is a precursor to LSD.

LSD, or lysergic acid diethylamide, was first synthesized by Swiss chemist, Albert Hofmann in 1938. Hofmann's work focused on isolating the chemical found in the ergot fungus. His goal was to use this chemical compound to treat medical conditions that targeted the respiratory and circulatory systems. While Hofmann discovered lysergic acid diethylamide at this time, it was only several years later that he learned of its hallucinogenic properties. In 1943, Hofmann accidentally ingested a small amount of LSD. Thereafter, he reported "extraordinary shapes with intense, kaleidoscopic colors" (*LSD*, 2017, June 14, para. 3). Intrigued, Hofmann decided to intentionally dose himself with the substance, experiencing the world's first LSD trip. In his book, *LSD-My Problem Child* (1980, p. 50), Hofmann described his experience as follows:

> Kaleidoscopic, fantastic images surged in on me, alternating, variegated, opening and then closing themselves in circles and spirals, exploding in colored fountains, rearranging and hybridizing themselves in constant flux. Every acoustic perception, such as the sound of a door handle or a passing automobile, became transformed into optical perception. Every sound generated a vividly changing image, with its own consistent form and color.

At the time, Hofmann was working for the pharmaceutical company, Sandoz. After reporting his experiences with LSD, Sandoz started animal trials to further research the chemical. However, LSD seemed to have none of the fantastical effects that Hofmann reported and only prompted a restlessness in the subjects (Miller, 2020). Interest in the compound waned until Hofmann was encouraged once again to re-look at the effects of LSD on the mind in the late 1940s. From there, LSD was marketed as Delysid, a pharmaceutical treatment for psychiatric disorders such as depression, anxiety, and other psychological trauma. Within less than thirty years, LSD had been studied and reported on in almost 10,000 scientific publications. It is currently the most studied psychedelic to date.

From there, LSD was infamously used in the 1950s MK-ULTRA project, a university-funded research study facilitated by the CIA (Central Intelligence Agency) in the US. The goal of the project was to investigate the use of LSD as a psychological weapon in the Cold War. Trials were conducted on volunteers as well as unwitting subjects. However, after almost a decade of research, the effects of the compound were found to be too unpredictable for use in the field.

LSD was famously used by a diverse range of people, from authors and artists to psychologists. Before 1968, LSD was still legal for use, allowing it to be used freely as well as in clinical trials. However, concern over the increasing recreational use of illegal substances prompted the US government to outlaw LSD.

Chemistry and Pharmacology

LSD is synthetically formulated from lysergic acid, a natural compound found in ergot fungi, one of the most common species being *claviceps purpurea*. Ergot is a fungal infection that occurs on rye grain as a result of the growth of these parasites. Lysergic acid is produced using ergometrine or ergometrine tartrate. Ergometrine is an ergot alkaloid or mycotoxin extracted from ergot fungi.

LSD is one of the most potent mind-altering substances and causes people to hallucinate and lose touch with reality. The molecular formula of LSD is C20H25N3O. It is structurally similar to serotonin, as well as naturally occurring lysergamides, which are chemicals found in various morning glory plants (*Ipomoea* spp.).

The term LSD is derived from the German name of the chemical, namely LysergSäureDiethylamid. Referred to also as lysergide (lysergic acid diethylamide), LSD falls within the same chemical family as psilocin and DMT. Commonly produced as a tasteless, odorless, and water-soluble salt, LSD is used recreationally as sheets of absorbent paper soaked in the substance, tablets, capsules, or solutions in water or alcohol.

While the mechanisms behind its effects on the brain are still not fully understood, studies suggest that the chemical binds to and activates the serotonin subtype 2-receptor (5-HT2), including both presynaptic 5-HT1A and 5-HT1B receptors, and 5-HT2 postsynaptic receptors. The collective agonist and partial antagonist effects on serotonin activity in the brain are thought to activate the release of glutamate in the thalamocortical terminals, which are the nerve fibers between the cerebral cortex and the thalamus. It is this reaction that causes the dissociation between the sensory relay centers and the cortical output to create hallucinogenic-type reactions. Historical records of the effects of ergot hint at its stimulation of the midbrain and sympathetic nervous system, which causes the somatic responses common with ingestion of LSD.

Because it has a strong binding affinity to the 5-HT2A receptors, it seems as though LSD artificially activates serotonin receptors, even though there is no serotonin being released. Normally, the body and brain work together to moderate awareness of the surroundings and filter the necessary information, which is then processed. However, with LSD, various neural pathways are stimulated, even without a causal stimulus. This encourages the brain to react to information that is technically not provided. For example, LSD can cause the brain to react as if light is falling on the retina, even though there is no external light stimulus, creating the hallucinogenic effects of the chemical.

The hallucinogenic effects of the substance are also related to the increased connectivity between the primary visual cortex and other brain areas. It is further believed that LSD redirects neural stimuli to the right side of the brain, which is the hemisphere related to creativity. Typically, LSD will reduce the integrity of normally functional neural circuits and stimulate connections between normally dissociated networks.

Further research on LSD suggests that it has other neurophysiological functions associated with the neurotransmitter dopamine, which creates feelings of well-being and motivation and controls mood, learning, concentration, sleep, and movement. This is important to note, as this can have an effect on the addictive potential of the chemical. Although both clinical studies and anecdotal data suggest there are no addictive qualities to LSD, the fact that it can act as a dopamine agonist means it can have the same reward-stimulating effects.

LSD may also interact with other postsynaptic neurotransmitters like adrenoreceptor subtypes, which modulate several physiological reactions like constriction and dilation of blood vessels, as well as muscle contraction and relaxation. LSD is processed in the liver and converted into inactive metabolites, which are expelled from the body in the urine.

Typical recreational doses of the chemical range from between 20 and 80 micrograms; however, the substance is already active on the user at 20 micrograms. Effects take place approximately 30 minutes after ingestion and may last for up to 12 hours. The dosage will determine the intensity and duration of the effects. Excessive doses of the chemical may result in what is colloquially called a "bad trip," which is characterized by panic, confusion, severe anxiety, and aggression. Incidences of serotonin syndrome have been recorded, as have significant physical reactions such as seizures, hyperthermia, rhabdomyolysis, hypertension, and tremors with severe LSD intoxication. Users who experience bad trips are treated with a benzodiazepine to induce sedation and prevent damage to the nervous system. There have been recorded cases of users experiencing unpredictable recurrences of the chemical's hallucinogenic effects for weeks or even months after use.

Therapeutic Applications of LSD

LSD is classified as a Schedule 1 drug under the United Nations 1971 Convention on Psychotropic Substances. Thus, it is still considered illegal in most countries across the world.

The first symptoms to appear once LSD is ingested are physical or somatic, namely dilated pupils, raised body temperature, and slightly elevated blood pressure. Thereafter, the user can experience sensory-perceptual changes like visual disruptions or illusory phenomena, and cross-sensory perception, where sounds present as colors or shapes. This is also known as synesthesia. The effects of LSD are divided into three categories:

- **Somatic**: tremor, weakness, dizziness, and paresthesias ("pins and needles")
- **Perceptual**: changes in hearing, altered visual sense

- **Psychic**: depersonalization, changes in mood, dreamlike states, and altered sense of time

Like most other hallucinogens, there are no recorded cases of physical or psychological dependence on the substance, and similarly, no recorded cases of death attributed to LSD. Additionally, there appear to be no significant withdrawal symptoms.

LSD can cause panic reactions in some users, which can last from several hours to two days and may trigger psychotic episodes. The reaction to the substance is highly dependent on the user's mental and emotional state, as well as the surroundings in which it is taken.

Clinical Studies and Research

Holze et al. (2020) developed a double-blind, randomized clinical study to investigate the effect of LSD microdosing to improve mood and cognition, as a reaction to the increased interest in the substance as a therapeutic enhancement. Their research showed that microdoses as low as 5 µg (microgram) did not elicit any subjective responses from the 23 participants. However, at 10 µg, participants showed generally positive effects, while participants reacted to the 20 µg dose with both positive and negative responses. Holze (2020) noted, however, that there are few clear guidelines on pharmacokinetic parameters (how the body reacts over the length of exposure to a substance) for the administration of LSD in therapy.

Recently, the FDA in the US granted a pharmaceutical biotech company with therapy designation for their LSD-based treatment for generalized anxiety disorder (GAD). A trial study of the salt form of LSD called MM210 showed statistically significant improvements in anxiety in participants who received the treatment as opposed to those who received the placebo. Participants reported this improvement after only one dose, and after several months, almost 50% of the study group noted that their anxiety was in remission (Terry, n.d., para. 6). In addition, participants reported only mild to moderate side effects of the dosage, including expected reactions such as nausea, vomiting, headache, dizziness, anxiety, and euphoria.

In another study by Holze (2022) 83 participants were involved in a double-blind, randomized, placebo-controlled study, where each subject received a single dose of LSD. Results showed that participants experienced subsequently increasing subjective, physiologic, and adverse effects with an increase in dosage. However, the increases in subjective positive effects were steeper in participants than subjective negative effects. In fact, for all but one of the four dose concentrations, namely 25 µg, 50 µg, 100 µg, and 200 µg, the majority of participants showed more than 50% positive effects (i.e. subjects experienced a subjectively positive effect from the chemical), namely 37%, 91%, 96%, and 91%, respectively. This is a significantly higher percentage than those experiencing more than 50% negative effects (0%, 9%, 27%, and 31%, respectively; Holze et al. 2022. p. 1).

Another double-blind, randomized, and placebo-controlled clinical study (Holze et al., 2022) aimed to investigate the effects of LSD-assisted therapy on patients struggling with anxiety. A sample group of forty-six participants was divided into two groups; one that received an LSD-infused dose and one that received a placebo. Thereafter, the participants were monitored for symptoms of anxiety for a period of 24 weeks. After another six weeks, the group that received the placebo first was given the LSD dose, and vice versa. This ensured that both groups of participants received both the placebo and the LSD dose. The results showed that both groups of participants experienced a marked decrease in feelings of anxiety, depression, and related psychiatric mood disorders when administered the LSD as opposed to the placebo. In fact, the group that first received the LSD dose showed a significantly more positive response than the group that received the placebo first. In addition, participants receiving the first dose of LSD showed a higher retention rate in the study than those who received the placebo.

This means that it is possible that the effects of the first dose of LSD actually overlapped with the second phase of the trial where those participants were due to receive the placebo. By the time the participants were required to receive the placebo, they were still reporting significantly lower anxiety scores. From there, participants showed an increasing improvement in anxiety scores, even without additional doses of LSD, as they moved on to the placebo phase.

Researchers suggest that this effect, called the carryover effect, is due to the long-lasting activity of LSD. In addition, participants who were already receiving treatment for anxiety before the trial showed a consistent, long-term improvement in their condition with the implementation of the LSD-assisted therapy. Interestingly, those who received the placebo first showed very little improvement in their anxiety symptoms even after the LSD dose in phase 2 of the trial, which in itself is concerning. Overall, this study specifically highlighted the long-term positive effects of LSD on reducing symptoms of depression and anxiety, in both patients that are physically healthy as well as those suffering from a life-threatening condition.

Liechti (2017) reviewed several clinical studies on LSD research and found that there is significant potential for the use of LSD in psychiatric therapy. One study showed that patients administered LSD had an impaired ability to recognize sad or fearful faces while increasing the patients' receptivity to suggestion, trust, and openness. Emotional empathy was increased, while the patients' amygdala responded less to fearful faces.

Experts suggest that the breakthrough designation of this form of LSD as a therapeutic measure is going to encourage the approval of LSD as a treatment for GAD in the near future, in essence making it legal for clinical use. Further clinical trials with larger focus groups are planned for the future, with a focus on testing the safety and effectiveness of LSD to treat GAD. As with other similar hallucinogenic-based treatment therapies, there is a three-pronged approach to the intervention:

1. patient education and preparation for the treatment
2. intervention or treatment itself, which will include a psychotherapist
3. post-treatment integration and retrospection session

Practical Considerations in Therapy

Historical records show the use of naturally occurring ergot in the early 1800s for inducing labor, although this practice declined after concerns of uterine spasms and danger to the child. First synthesized by Dr Albert Hofmann in Switzerland in 1938, in his quest to develop medication from the compounds in ergot, the chemist unintentionally ingested the substance while processing it for purification and crystallization. Based on his experiences of a distortion of time, and visions of vivid highly-colored images, Dr Hofmann was prompted to further investigate the reason behind his reactions to the chemical. From there, he conducted various self-experiments on the ergot alkaloid in small quantities of 0.25 mg.

In recent years, interest has grown in the use of LSD in treating depression, anxiety, cluster headaches, and substance abuse. In addition, the use of LSD to microdose (very low doses at intervals of 2-5 days) for various ailments has increased, specifically to improve moods and cognition.

Most importantly, studies are underway to test the effectiveness of LSD to treat anxiety disorders, both alone and in conjunction with psychotherapy. Many experts suggest that psychotherapy, or at least the presence of a trained therapist, is important, especially in the event that the participant has a negative experience with the chemical. In addition, it is imperative to conduct the experience in a highly controlled environment, one in which participants are encouraged to focus on their inner experiences rather than their external environment. In order for participants to receive the most benefit from the therapy, they need to feel safe, especially as the LSD can create a disconcerting experience for some.

Overall, clinical trials have shown remarkably overall positive responses to LSD, specifically in patients suffering from various psychiatric disorders such as depression and anxiety. In general, patients experience little to no adverse effects, and the overall positive effect is long-term. This bodes well for the future potential of LSD to be used to treat these and other psychiatric disorders.

However, while various small-scale clinical studies have highlighted the potential for LSD to assist in treating patients with depression, it is recommended that larger studies be conducted to validate these benefits. Also, what is unclear at present, is whether the reported positive effects of LSD on patients is a result of psychological or pharmacological responses. The Holze et al. (2022) study discussed above, where participants who first received a dose of LSD reported long-term positive effects as opposed to those who received the placebo first, suggests that the results of the LSD may be a consequence of the context of the study rather than the actual effect of LSD. There is still uncertainty as to the therapeutic actions of LSD. This needs further study through modern clinical research.

McCulloch et al. (2024) highlight several knowledge gaps in the clinical studies on the effects of LSD-assisted therapy on the treatment of psychiatric disorders:

- There is little research currently into the clinical pharmacology of LSD.
- Little is known about the effects of age, sex differences, and pharmacogenetics (how genetics affects the patient's response to drugs), amongst other moderating elements, on the effectiveness of LSD in treating psychiatric disorders.
- More precise physiological and psychological measures to test the effects of LSD need to be implemented across studies to ensure consistency in results comparisons.
- Various personality traits such as neuroticism or the potential to surrender may influence the outcome of the treatment and need to be taken into account when analyzing the results.

- Further research needs to be done on the length of the LSD-assisted therapy required to achieve long-lasting therapeutic effects.
- It is suggested that the expectations of the patients prior to the treatment, as well as prior experience with hallucinogenic substances, need to be taken into account, as these may influence the outcome of the study.
- More research needs to be done on the safety of administering LSD-assisted treatment in conjunction with standard antidepressant medication, or the tapering of this medication.
- The setting and environment, as well as additional cognitive therapy provided during the LSD-assisted treatments, need to be consistent across trials to ensure results that can be reliably compared.
- In order for LSD to be approved as a treatment for psychiatric disorders, especially as it is currently illegal in most countries, clinical trials need to be well-designed, randomized, and placebo-controlled. The benefits of the treatment versus the risks need to be clearly and statistically demonstrated to prove its safety and reliability.

MDMA

The first recorded medical use of MDMA, originally referred to as methylsafrylaminc, was in 1912, when it was marketed by German pharmaceutical company, Merck, as a parent compound to synthesize medications for use in controlling bleeding. Because it was simply a precursor to other drugs, it was not tested on humans for its psychoactive properties. However, pharmacological research was done in non-human trials at Merck in 1927 and 1959.

Although the substance had not received approval from the US FDA and had not been used in formal clinical trials, several psychiatrists used it to treat psychiatric disorders throughout the late 1970s and early 1980s. Because it encouraged empathy in patients, medical professionals believed it opened up channels of communication between therapists and their patients. This, they understood, allowed patients to delve into their past issues and trauma, and assisted them in achieving greater insight into their challenges. At the same time, MDMA was gaining popularity as a party drug, where it was used extensively at nightclubs and dance parties or raves.

The first notable evidence of the drug being used in a recreational setting was in Chicago in 1970. It is believed that the substance was synthesized in an attempt to recreate the more popular drug, MDA, which was at that point, illegal for use and possession under the Controlled Substances Act. MDA is a synthetic psychedelic that is similar in structure to MDMA but is believed to have greater psychedelic effects, causing significant visual hallucinations. Both substances cause distorted sensory experiences, an elevated mood, and an increase in energy and pleasure. Both substances are popular as party drugs and are associated with the club scene and music festivals. They are popular for the combination of both stimulant and psychedelic effects. Because MDMA has greater physiological effects, such as increases in heart rate and blood pressure, it is considered more dangerous than its counterpart. However, MDA causes more severe jaw clenching and tension than MDMA.

The National Survey on Drug Abuse and Health has noted that more than 18 million US adults have tried MDMA at least once.

Chemistry and Pharmacology

MDMA, also commonly known as ecstasy or molly, is a synthetic, psychoactive substance that was developed for the first time in 1912. It is listed as a Schedule 1 illegal substance under the United Nations 1971 Convention on Psychotropic Substances. It was patented by Merck Pharmaceuticals in 1914 as a precursor to other synthetic chemicals, especially to be developed and marketed as appetite suppressants. Because of its increased use as a recreational drug in the 1980s, it was scheduled under emergency as a Schedule 1 drug by the Drug Enforcement Administration in 1985. While it was once used relatively commonly as a form of treatment in psychiatric therapy in the 1970s and 1980s, its listing as an illegal substance has decreased its application. However, in the early 1990s, the FDA approved MDMA for use in the first human trial to investigate the substance in psychedelic-assisted therapy.

It forms part of the ring-substituted phenethylamine group, which includes methamphetamine and mescaline. MDMA is a chemical derivative of amphetamine, which is a stimulant drug that increases the rate at which information is passed along the neural circuits of the brain. Because of its chemical structure, it shares common characteristics with both hallucinogens, like LSD, as well as stimulants, like amphetamines. It provides a combination of sensual, empathetic, and stimulant effects.

The phenethylamine molecule in MDMA consists of an amino group, which is attached to the phenyl ring by an ethyl chain. Phenethylamine, which is an organic compound that occurs naturally as well as in synthetic drugs, is the foundation of many pharmacological substances like adrenaline, serotonin, and dopamine. It is also the backbone of several psychoactive substances, such as

MDMA. In a ring-substituted phenethylamine group, the phenethylamine has one or several substituents attached to the phenyl ring. The chemical and biological properties and activities of ring-substituted phenethylamines can be altered by changing those substituents.

Phenethylamines without ring substitution generally have a stimulant effect; however, phenethylamines with ring substitution, like MDMA, show changes in their pharmacological properties. The effects of MDMA are generally feelings of euphoria, mild central nervous system stimulation, and an increase in sensory awareness and perception. This is different from the effects of other related stimulants like amphetamine. As with other amphetamines, MDMA stimulates the production of neurotransmitters like dopamine, serotonin, and norepinephrine. MDMA works by actively binding to serotonin transporters and acutely increasing the availability of serotonin in the brain.

The coordinated release of dopamine and norepinephrine can cause sleep deprivation and disturbed sleep, both of which are notorious side effects of MDMA use. Clinical animal trials have shown that MDMA causes the release of serotonin by effecting leakage of the neurotransmitter out of the cell, rather than the normal process of the activation of a nerve impulse. This suggests damage to serotonin nerve terminals, which may cause MDMA's neurotoxic properties.

MDMA affects the brain by increasing activity in the superior frontal cortex and increasing connections between the amygdala and hippocampus. It also reduces the amygdala's response to negative emotional stimuli. The release of serotonin, dopamine, and norepinephrine stimulates the secretion of other hormones such as oxytocin. Oxytocin is colloquially referred to as the love or bonding hormone, as it is released during positive social interactions such as hugging. It encourages a reduction in stress and anxiety and promotes trust and empathy.

MDMA is different from traditional amphetamines in its ability to act as an empathogenic or entactogenic substance. This is because it enhances feelings of empathy, compassion, and interpersonal connections (empathogenic), as well as emotional openness and tactile sensations (entactogenic). This gives MDMA its socializing effects. Research has shown that MDMA also reduces the user's neurophysiological fear reaction to emotional threats and increases introspection.

The ability of MDMA to increase communication between users, as well as social bonding, makes it an ideal substance for use in assisted therapy to treat psychiatric disorders such as anxiety and PTSD. Studies show that more than three percent of adults in the US suffer from symptoms of PTSD, with severe cases sometimes resulting in related psychiatric disorders such as substance abuse, addiction, and suicidal ideation (Sreenivas, 2023). Unfortunately, while there are various treatment options available, such as cognitive behavioral therapy, almost 58% of sufferers still present with symptoms of PTSD after treatment (Sreenivas, 2023).

MDMA is usually synthesized using four organic compounds, namely safrole, isosafrole, piperonal, and 3,4-methylenedioxyphenyl-2-propanone. These principal precursors are all listed under the United Nations 1998 Convention Against Illicit Traffic in Narcotics and Psychotropic Substances.

Once a dose is administered, the maximum plasma concentration is reached within two hours, whereafter the effects diminish over six to seven hours. Animal trials have shown that MDMA can cause neurotoxicity (damage to the neurons and glial cells of the brain) and a decrease in the brain's production of serotonin. Use in humans has been related to cognitive impairment, such as memory loss, trouble concentrating, and issues with following instructions and solving problems.

Therapeutic Applications of MDMA

Currently, there are few clinical studies that highlight the use of MDMA in substance-assisted therapy. However, clinical trials are steadily increasing, especially with positive results from smaller studies that have looked at the use of MDMA for the treatment of PTSD, notably for its ability to reduce anxiety in users.

Because of the ability of MDMA to increase the production of serotonin in the brain, it has been earmarked as a potential treatment for certain psychiatric conditions, such as PTSD. It has been classified as a breakthrough therapy by the Food and Drug Administration (FDA). This classification is set to encourage further research and development on its use in therapy. In addition, patients with life-threatening, treatment-resistant PTSD who are unable to join clinical trials are now able to attend MDMA therapy.

Researchers suggest that the activity of MDMA in the brain that reduces defensiveness and anxiety can assist patients in accessing traumatic and uncomfortable memories. This can allow them to process and deal with these memories under calm, less stressful conditions. The empathogenic nature of the substance can also encourage the patient to open up to their therapist and promote more productive sessions.

Clinical-use MDMA has a slightly different chemical structure to the street drug (ecstasy), as it is pharmaceutical-grade with known quantities of the chemical and no other additives. A regulated dose of the substances is administered to patients in a controlled environment under the guidance of professionals. Generally, a therapy-assisted treatment session with MDMA involves the patient receiving a 120 mg dose of the substance. If required, an additional dose is administered approximately two hours after the first dose. The session, and the effects of the substance, last for around eight hours, during which the patients engage with a pair of trained psychotherapists to work through past trauma. Typically, the patient participates in two to three similar sessions over twelve weeks.

At this stage, MDMA is being used in therapy-assisted treatment of PTSD, anorexia nervosa, binge-eating disorder, and anxiety disorders. The benefits of this therapy suggest that, unlike other conventional treatments for PTSD, MDMA-assisted treatments provide relief from symptoms after a single course of treatment only. Various trials also show that up to 75% of participants in this research have a complete disappearance of symptoms, and no longer report that they are suffering from PTSD a year after their treatment (Yehuda, n.d.).

Clinical Studies and Research

In a recent clinical study, Wolfson (2020) implemented a double-blind, randomized study on the effect of MDMA on life-threatening cases of psychiatric illnesses, like depression, anxiety, and existential distress. Participants received either a placebo or a 125 mg dose of MDMA, along with two eight-hour psychotherapy sessions. Results showed that participants who received the treatment had a greater average reduction in anxiety than the control group, although this difference was not highly significant. None of the participants experienced significantly marked negative effects, although there were reports of dry mouth, headache, thirst, jaw-clenching, fatigue, and insomnia. The only significantly different physiological response between the treatment and control group was elevated body temperature; however, these effects did not require medical intervention. Follow-up visits at six months and one year showed a stable improvement in anxiety and depression, which emphasizes the long-term benefits of MDMA-assisted psychotherapy.

Later, a randomized, placebo-controlled, double-blind study by Mitchell (2021) looked at the use of MDMA-assisted therapy to treat severe PTSD. Participants were also experiencing several comorbidities, such as depression, dissociation, substance abuse and addiction, and childhood trauma. Before embarking on the treatment, participants were weaned off current medications. Treatment was provided in conjunction with therapy sessions from a medical professional. Results showed that participants who received the treatment had a significant decrease in impairments caused by pre-existing psychiatric disorders. Further, participants did not report any incidents of suicidal ideation or potential for abuse of the substance.

In 2023, Mitchell and colleagues rolled out phase three of their investigation into the use of MDMA to treat moderate to severe PTSD. Because only 35-47% of PTSD sufferers benefit from treatment with SSRIs, the study hoped to research more effective ways to treat it. Various phase two trials positively indicated that there was an acceptable risk-to-benefit tradeoff in those with PTSD who were treated with MDMA (Mitchell et al., 2023). In other words, the benefits of MDMA-assisted treatment for PTSD outweighed the risks. Results of the third phase of the project, which showcased an enthno-racially diverse sample, showed that participants treated with MDMA-assisted therapy had a significant decrease in their symptoms when compared with those who received therapy only. In addition, behavioral impairments related to PTSD were significantly mitigated with MDMA-assisted therapy. It was noted that 98% of participants experienced at least one treatment-emergent adverse event (TEAE) (Mitchell et al., 2023), which included nausea, decreased appetite, feelings of heat or cold, dry mouth, restlessness, or tremors. However, less than 9% experienced a severe TEAE (Mitchell et al., 2023). Overall, the study revealed that over 86% of participants with moderate to severe PTSD who received the treatment experienced relief from symptoms compared with those in the control group (Mitchell et al., 2023). In addition, researchers were pleased to note that no major safety issues were highlighted throughout the study; any TEAEs experienced were consistent with those generally occurring with MDMA use.

Clinical trials have suggested the most effective use of the substance in therapy-assisted treatment is to administer it over three sessions of eight hours long that take place every 3-5 weeks. In conjunction, patients are required to attend twelve sessions with a therapist. In general, studies have shown that MDMA-assisted therapy is effective at providing a sustained reduction in symptoms of PTSD in patients who experience treatment-resistance disorders. Its effectiveness has largely been attributed to the mechanisms of the amygdalar serotonergic functions, which control fear-based reactions, especially those that contribute to sustained PTSD. MDMA is believed to facilitate the processing of previously intractable fear-related memories (Mitchell et al., 2021).

Practical Considerations in Therapy

The use of MDMA in assisted therapy for the treatment of PTSD is gaining traction, as many trials highlight its effectiveness in controlling or even removing symptoms after three sessions. Because of this, the FDA has recently granted MDMA expanded access status. This classification allows patients with severe or life-threatening conditions to be given access to these investigational drugs outside of clinical trials. In addition, treatment with MDMA is often more cost-effective than other traditional methods. Added benefits include the fact that MDMA does not have the common hallucinatory effects associated with other psychedelics.

However, there are still several concerns with the use of MDMA in the treatment of psychiatric disorders. High doses of MDMA can cause a rapid increase in body temperature, which can lead to cardiovascular and kidney system failure and muscle disintegration. Other symptoms of toxicity include hallucinations, hypertension, anxiety, agitation, rapid breathing, increased heart rate, and clenching or grinding of the teeth. Fatalities linked to the use of MDMA have been reported, although the potential for it to be toxic is dependent on individual susceptibility and how it is administered. Excessive use of MDMA can also result in serotonin syndrome, especially if taken in conjunction with other illicit serotonergic substances or medications like SSRIs. If used regularly, MDMA can also cause panic disorders, anxiety, depression, paranoia, and insomnia.

The neurotoxic effect of MDMA is related closely to its action on mitochondria, which are the powerhouses of the body's cells and generate the energy required to activate biochemical reactions. This impairment of the activity of mitochondria is especially significant in the brain, where MDMA impairs the ability of mitochondria to produce energy and detoxify cells. In brain areas with high concentrations of serotonin-producing neurons, the effect of MDMA-induced neurotoxicity is worse.

Due to its euphoric and stimulating effects, MDMA has a high risk of dependence and addiction. This is related to its similarity and link with dopamine and serotonin, both of which are associated with substance abuse, addiction, and impulsivity. However, there is no research that confirms whether MDMA is addictive or not. Anecdotal evidence suggests that regular users of MDMA may experience withdrawal symptoms, as well as cravings and tolerance to the substance.

As it stands, many studies on the effect of MDMA on the brain often don't take into account pre-existing conditions and other drugs used in conjunction. Certain clinical trials have suggested that patients who have recently used SSRIs may not respond as favorably to MDMA-assisted therapy. This brings into question the effectiveness of this type of treatment for those with a long-term history of SSRI use. These concerns require that further peer-reviewed studies be done to determine the safety of using MDMA in the treatment of PTSD and other psychiatric disorders, especially in patients with comorbidities and those on existing medication.

Currently, beyond its status as an expanded access drug, the FDA is yet to classify MDMA as an available treatment option. The use of MDMA in the treatment of PTSD can only take place in an approved and regulated setting with trained professionals. However, its success in several clinical trials shows promise for its effectiveness in the treatment of various psychiatric disorders.

KETAMINE

In the 1950s, the American pharmaceutical company, Parke-Davis, was working on developing the ideal anesthetic with analgesic properties. In 1956, the lab was investigating the family of cyclohexylamines, which led to the discovery of a synthetic compound called PCP (*N*-(1-phenyl-cyclohexyl)-piperidine) or phencyclidine. From there, various researchers started animal clinical trials to study the effects of PCP as an anesthetic. However, while animal subjects could be operated on with no evidence of pain, muscle relaxation was poor, with a loss of the body's ability to manage blood pressure. In addition, it appeared as though there was no loss of consciousness.

Researchers continued to study the chemical compound, this time trialing it on human subjects as an anesthetic under the name Sernyl. Clinical trials showed the PCP worked effectively as an anesthetic, producing a cataleptic state (where the patient is not responsive to external stimuli, and muscles show a lack of rigidity). In addition to catalepsy, PCP was able to induce amnesia and analgesia (the inability to feel pain). However, the use of PCP as an anesthetic caused certain side effects; several patients showed extreme excitation on administration of the substance, while more than 15% were unmanageable in post-operative care. In addition, several patients took between 3 and 18 hours to recover from the anesthetic (Mion, 2017). Excitation is defined as an increase in brain activity on administration of an anesthetic drug, at the same time causing a loss of consciousness.

PCP appeared ideal as an analgesic; it didn't have significantly negative impacts on the cardiovascular or respiratory systems, nor did it depress pharyngeal or laryngeal reflexes. However, it was limited by excitation that persisted, often for more than 12 hours after administration. Further research suggested that PCP was not suitable as an anesthetic in humans.

Years later, Calvin Stevens, an American chemist, synthesized a short-acting anesthetic that was selected for human trials. Described as a cataleptic compound with anesthetic and analgesic properties without the hypnotic properties of PCP, the substance was called ketamine. In the late 1960s and early 1970s, ketamine was used in human clinical trials across the world. While providing excellent analgesic and anesthetic properties, concern was raised over its shorter duration of action, as well as its lower potency levels in comparison with PCP. Further to that, patients reported unpleasant hallucinations in conjunction with the substance, all of which meant its use was later halted.

Ketamine was patented for use predominantly as a veterinary anesthetic in 1963. In 1970, ketamine was prescribed for human use under the name Ketalar, after which it was officially approved by the FDA. Because of its wide margin of safety, its ability to improve respiratory function, and its stimulation of the nervous system (reduced fatigue and heightened alertness), Ketalar was used as a field anesthetic for American soldiers during the Vietnam War.

Following an increase in ketamine abuse during the Vietnam War, as well as its abuse amongst the general public, specifically along the US East Coast, it was later categorized as a Class 3 substance under the Controlled Substances Act in 1999.

Chemistry and Pharmacology

These days, ketamine is a form of anesthetic generally used during surgery on animals. Its chemical structure is 2-(2-chlorophenyl)-2-(methylamino) cyclohexanone. Ketamine is soluble in both lipids and water. Its lipophilicity (fat-solubility) means it can be administered orally, nasally, intramuscularly, intravenously, and subcutaneously. Because it is so highly lipophilic, ketamine passes easily through cell membranes, which causes it to be short-acting. In other words, it has a rapid onset of action. It is a non-barbiturate that causes anesthetic effects but preserves normal swallowing or gag reflexes, as well as respiratory drive and reflexes. This dissociative anesthetic distributes quickly through the body, making it beneficial as pain treatment, and as a way to induce anesthesia. It is this property that also causes users to feel disconnected.

There is still a varying degree of uncertainty regarding the exact mechanisms of ketamine in the brain. It is generally thought that ketamine works as an antagonist of a type of glutamate receptor called the N-methyl-D-aspartame (NMDA) receptor. As discussed in previous chapters, glutamate works to encourage synaptic transmissions, which is essential for various cognitive processes like reasoning, memory, and learning. Ketamine binds to the NMDA receptors, blocking glutamate from exerting its excitatory effects on neurons. Normal information transmission in the brain is disrupted, affecting mood regulation, perception, and cognition.

While ketamine blocks NMDA receptors, it also promotes the release of other neurotransmitters like dopamine and serotonin elsewhere in the brain. Other glutamate subtype receptors, like the alpha-amino-3-hydroxy-5-methyl-4-isoxazolepropionic acid (AMPA) receptors are activated by ketamine, which encourages synaptic plasticity, where synapses are strengthened or weakened. Ketamine can enhance neuroplasticity in a process that allows users to break out of depressive states. Researchers suggest that ketamine allows new neural pathways to form and restricts neuronal atrophy and synaptic defects.

The Default Mode Network (DMN) is a system of brain networks, most often engaged when the brain is at rest or when daydreaming, introspection, or engaging in self-referential thinking (relating external stimuli to oneself based on past experiences and memories). The DMN is strengthened with frequent use of specific neural circuits. Someone with a positive outlook on life will stimulate those networks associated with optimism, further encouraging them. However, for those suffering from depression, a negative viewpoint is cemented by the frequent use of specific neural pathways involved with unhelpful thoughts. In other words, a depressed person's DMN will be fixed in a system of negative thought patterns. Ketamine is thought to disrupt negative DMNs in sufferers of depression, allowing them to rebuild positive neural networks and promote positive thinking. That is why medical professionals suggest ketamine be used as a supplement to other forms of behavioral therapy, in what is called ketamine-assisted psychotherapy.

Recent research also shows that ketamine may affect the lateral habenula, an area of the brain that processes emotional and behavioral responses to negative stimuli like stress, anxiety, and pain. It is also involved in modulating the sleep-wake cycle. The lateral habenula is often over-reactive in those suffering from depression. Ketamine is thought to reduce the hyperactivity of the lateral habenula, having an anti-depressant effect.

In essence, neural imagery suggests that ketamine has both an inhibitory as well as an excitatory effect on the brain, both dampening activity while awakening the brain. In a study on the effects of ketamine on the brain, neuroscientists, Takaa Hensch and Michela Fagiolini, found that ketamine affects NMDA receptors on specific neural cells called parvalbumin neurons. These neurons work to depress the activity of surrounding brain circuitry, so when ketamine blocks glutamate in these neurons, their inhibitory effects are eliminated, creating excitatory effects in that part of the brain. In this way, the sensitivity of the brain to sensory inputs is increased, decreasing symptoms of depression. The study concludes that the parvalbumin neurons are the main target of low doses of ketamine (Picard et al., 2019).

Therapeutic Applications of Ketamine

Ketamine was first synthesized in the 1960s as a derivative of PCP (phencyclidine). It has a shorter half-life and fewer side effects than the original compound. However, like PCP and other psychedelics, it can cause hallucinations. It is now used as a dissociative anesthetic, specifically for medical procedures that require general anesthesia (a sleep-like state) and not total muscle relaxation. Doctors use it when short-term sedation is required, usually in an emergency setting, to address issues such as joint dislocations and fractures.

Ketamine is rapidly metabolized, mainly by the liver. Because of this, and its rapid excretion, it facilitates a smooth and gentle recovery from sedation or anesthesia. Ketamine is also sometimes used in low doses to treat severe pain resulting from fractures, trauma, or back and abdominal issues. It is approved by the Federal Drug Administration (FDA) for use in skin grafts, orthopedic procedures, and any diagnostic procedures performed on the nose, eyes, ears, and throat.

It is a Class 3 controlled substance, which means it has a "moderate to low potential for physical and psychological dependence" (*Drug scheduling*, n.d., para. 5). Its dissociative properties cause a disconnection between cognitive processes such as thinking and reasoning, and external stimuli. This produces an analgesic effect, and causes "amnesia, relative immobility, and an altered state of consciousness" (*Ketamine*, n.d. para. 1). Ketamine can take effect within three to five minutes after administration, peaks quickly, and has a duration of around twenty to thirty minutes. These characteristics are what make it so useful in a pre-hospital setting for sedation of patients who require examination by emergency response teams. It has no significant impact on blood pressure, ventilation, or airway reflexes, as opposed to similar sedatives.

It is used primarily for inducing and maintaining anesthesia but is also used off-label in several other situations, such as to relieve extreme agitation, to stop seizures, as an anti-depressant, and as a pain reliever. Any off-label medication is that which is used in a medical setting for conditions it has not yet been approved to treat.

In certain situations, emergency response teams use ketamine to subdue patients. Studies have shown that intramuscular doses of between 3-5 mg/kg (milligrams per kilogram) of ketamine are considered safe to administer to patients prior to evaluation by emergency management services in a pre-hospital setting. However, data shows that there are several potential complications that can occur in a pre-hospital situation, which include hypoxia, raised blood pressure, and obstruction of the airway. These reports are conflicting across various institutions: Experts such as anesthesiologists and toxicologists suggest that paramedics and other emergency response personnel may not be familiar with the dissociative effects of ketamine, which causes them to misinterpret the effects on the patient.

Clinical Studies and Research

Various clinical studies have looked at the potential for ketamine to be used for treatment-resistant psychiatric conditions such as depression and anxiety. Its effectiveness as a rapid-onset treatment for mood and anxiety disorders means it is now being used as an off-label treatment for these and other psychiatric conditions. Because of its addictive quality, concern has been raised by mental health professionals about its use for treating depressive disorders. For an anti-anxiety or anti-depressant substance to be labeled effective, it needs to significantly reduce symptoms of these disorders without concerns of safety and liability issues.

A study conducted in 2018 investigated the use of ketamine as a treatment for social anxiety disorder (SAD). SAD is a psychiatric disorder that causes the sufferer to experience feelings of social inadequacy, as well as a persistent fear of social interactions. Approximately 12% of adults in the US suffer from SAD, with approximately two-thirds of sufferers not receiving any relief from symptoms with the use of conventional medical treatments (Taylor et al., 2018, 325). The symptoms of SAD result in a severe decrease in quality of life, as well as distress, and other comorbidities. Taylor (2018) conducted a randomized, placebo-controlled, double-blind clinical trial with 18 adults diagnosed with SAD. Participants received either an intravenous administration of ketamine or a placebo saline solution. Participants were required to report on their symptoms of social anxiety throughout the trial. Results showed that participants who received the treatment had significantly lower symptoms of social anxiety after the trial as opposed to those who received the placebo.

Other clinical trials have shown that ketamine has a significant and rapid effect on decreasing symptoms of depression, even after it has been metabolized and excreted from the body. However, the reduction in depressive symptoms is not sustainable, with patients showing a rate of relapse of up to 73% at only a month post-treatment (Zhang et. al., 2016, 2). Research also suggests that ketamine has a significant anxiolytic effect (anti-anxiety), evident by the reduction in anxiety in patients with major depressive disorder with anxiety as a comorbidity. Further, randomized, placebo-controlled trials have reported the potential benefits of ketamine in treating obsessive-compulsive disorder (OCD) and PTSD.

In future, clinical studies need to address the potential for participants to identify that they have received the treatment and not the placebo, which may have significant effects on the results by creating bias. Studies also suggest that more research needs to be done on the specific dose of ketamine required to successfully treat anxiety.

Practical Considerations in Therapy

The potential for ketamine to relieve severe pain, cause dissociation, and produce hallucinations causes concern for abuse and inappropriate use of the substance, potentially resulting in addiction. This means it requires careful regulation and guidelines for use. Generally, ketamine causes psychological rather than physical dependence, with the majority of users reporting a craving for the substance when no longer taking it. Therefore, medical professionals recommend that patients with a risk or history of psychotic disorders, such as schizophrenia and bipolar disorder, should not use ketamine. Because of its rapid-onset sedation effect and potential to cause memory loss, ketamine is sometimes abused as a date-rape drug.

If administered at prescribed doses, ketamine is relatively safe, causing moderate side effects such as dizziness, vomiting, nausea, confusion, and double vision. Nonetheless, there are several risks associated with use, including the following:

- unstable heart rate and blood pressure
- shallow or slow breathing (hypoventilation)
- confusion and agitation
- liver dysfunction
- cognitive deficits

There are also various symptoms of withdrawal associated with the use of ketamine, including anxiety, impaired judgment, disorientation, and pain. The dissociative anesthetic qualities of ketamine that cause it to induce sedation may also result in incapacitation and memory loss if not administered correctly and safely. It can also negatively interact with alcohol. Ketamine should not be taken in conjunction with other substances such as benzodiazepines, opioids, and other anti-anxiety medication, as this can induce deep sedation, coma, and even death. There is a potential for ketamine overdose, which can result in severely impaired alertness, as well as death. Other symptoms of overdose can include high blood pressure, increased heart rate, muscle rigidity, and unconsciousness.

In order to prevent potentially adverse reactions to the substance, the FDA has recommended that the patient receiving a ketamine dose be restricted from any verbal, visual, or tactile stimulation during treatment. However, a comprehensive review of clinical studies on the regulated use of ketamine in a professional environment shows that treatments are generally safe and well-tolerated (Zhang et al., 2016). Considerations of future use suggest that more research needs to be done on the risks of administering multiple ketamine doses over long periods of time. This is because the relatively short half-life of the substance renders it ineffective after a certain period, with depressive and anxiety symptoms returning shortly after administration. In addition, medical professionals caution that multiple doses of ketamine, even under supervision, can create drug dependency in patients. Anecdotal evidence from the off-label use of ketamine to treat psychiatric disorders has reported patients requesting additional infusions to counter the withdrawal or mood crashes experienced after a single infusion, even in a controlled setting.

Ketamine is highly abused in some countries, creating serious public health concerns and decreasing tolerance for the use of the substance as an off-label treatment for medical conditions. These concerns have restricted the use of ketamine for treating psychiatric disorders and have also limited clinical studies. Because it is rapidly metabolized and excreted from the body, its effects are short-lived, which means there is a need for multiple doses, further increasing the risk of the development of dependency. Overall, ketamine doesn't show any significant benefits in treating depression and anxiety when compared with conventional medications, such as SSRIs and other drugs. This is especially significant when these conventional medications are taken in conjunction with cognitive behavioral therapy and other psychiatric treatments.

Zhang et al. (2016) suggest that ketamine should not be used to address depression, specifically because it can be misused at higher doses, and because current safety data is only based on small clinical trials and short periods of observation. There are currently few, if no, studies that indicate that ketamine has long-term benefits for the treatment of psychiatric disorders. Further, more robust, long-term clinical trials need to be conducted before ketamine treatment can be recommended safely. Clinicians who prescribe ketamine off-label need to carefully consider the ethical implications of the treatment, as well as observe the substance's dependency potential. Ketamine should only be administered if patients have tried all other standard anti-depressants and anti-anxiety medications with no relief from severe symptoms.

AYAHUASCA AND DMT (N,N-DIMETHYLTRYPTAMINE)

Indigenous tribes from all over the Amazon, from Bolivia and Columbia, and Peru to Venezuela, have used psychedelic compounds for centuries to enhance and integrate spiritual experiences. Ayahuasca is well-known for its use in religious, shamanic, and healing rituals, allowing the user to commune with ancestors and other entities in the spiritual world. While the exact origin of its use has been lost to history, remnants of the compound found infiltrating clay pottery take its use back thousands of years.

European colonizers first encountered the brew between the 16th and 19th Centuries, observing and documenting its use in spiritual ceremonies. In 1851, the English botanist, Richard Spruce, described the caapi vine as one of the predominant ingredients in the tea. Almost one hundred years later, more comprehensive work was done on the use of ayahuasca, specifically by Richard Evans Schultes, an American ethnobotanist and conservationist. Schultes already had a history of interest in natural psychedelic compounds, having spent time experiencing and researching substances such as peyote and psilocybin. In fact, Schultes was instrumental in the scientific discovery of psilocybin.

From there the interest in psychedelics, ayahuasca and psilocybin in particular, grew among the American public, with many writers, researchers, and explorers traveling to Central and South America to partake in these spiritual ceremonies. In 1968, researchers Jan-Erik Lindgren and Laurent Rivier visited Peru, where they noted that many of the indigenous tribes of the area actively partook in ayahuasca. However, only the shamans and healers were allowed to ingest it, and by doing so would assist tribe members with issues and illnesses by engaging with deities, ancestors, and entities in the spiritual world.

Nonetheless, various observations by explorers and researchers noted that in some tribes, both men and women, even those who weren't healers, would drink the brew, either to bond with one another through these experiences or to learn more about nature and its spirits.

Recreational use was prohibited in 1971, when the United Nations Conventions on Psychotropic Substances categorized it as a Schedule 1 substance, making it illegal to use, possess, and distribute.

Chemistry and Pharmacology

Ayahuasca is a tea infusion made from the caapi and chacrona plants that occur in the Amazon. The bark of the *banisteriopsis caapi* vine contains beta-carboline harmala alkaloids, harmine, and harmaline, which have antioxidant, anti-viral, and anti-microbial properties. *Psychotria viridis* (chacruna) a perennial shrub, contains within its leaves the hallucinogen DMT. Ayahuasca has been used traditionally for thousands of years for healing, religious ceremonies, rites of initiation, and spiritual rituals by South American shamans. The consumption of ayahuasca by shamans and their patients would first cause severe vomiting and sometimes diarrhea, after which the user would experience visions and spiritual insights. Typically, a shaman or curandero (traditional healer) brews the torn leaves of the chacrona with stalks from the caapi to create a tea.

Consumption of the sacred tea is done in a ceremony, presided over by the shaman. The shaman guides and assists the participant throughout the ritual, especially because the experience involves extended periods of altered states of consciousness. Shamans also monitor the participants for ill effects and to ensure safety throughout. The ceremony usually takes place at night and lasts until the effects of the brew wear off. Generally, users reach a peak experience between half an hour and an hour after taking the tea, after which the trip bottoms out after two to six hours.

DMT is structurally similar to serotonin and chemically resembles other indolealkylamines, such as melatonin and psilocybin. Indolealkylamines are chemical and structural analogs of 5-hydroxytryptamine (5-HT), which is the neurotransmitter serotonin. DMT is also found in legumes, frogs, grasses, marine sponges, fungi, and tunicates. As with other psychedelics, DMT is a serotonin receptor agonist. It is believed to activate the 5-HT2A serotonin receptor subtypes, which affect the release of neurotransmitters like glutamate and dopamine. It has a low bioavailability, which means it is readily metabolized in the body and excreted. Studies suggest that DMT, like other psychedelics, excites inhibitory interneurons in the visual cortex, disrupting the normal transmission of information and causing visual abnormalities, such as halos, waves, and trails.

DMT has also been shown to prevent degeneration in the brain by activating a protein called the sigma-1 receptor (SIG-1R). SIG-1R works to modulate various mechanisms involved in neurodegeneration, which is the damage of brain cells. These receptors coordinate repair in the brain by regulating synaptic plasticity and neuronal growth, as well as producing anti-inflammatory effects. DMT is also thought to increase the brain-derived neurotrophic factor (BDNF), another protein that facilitates healthy brain functioning by modulating inflammatory responses, encouraging neuroplasticity, and providing anti-oxidative effects. Although this is ongoing research, there is increasing evidence that points to the benefits of DMT to healthy brain functioning. This has prompted further research into the use of ayahuasca for the treatment of degenerative brain disorders like Alzheimer's disease.

Beta-carbolines are a type of indole alkaloid. Indole alkaloids are chemical compounds that are found naturally in certain plants, such as fungi and trees. They consist of a hexagonally shaped ring of carbon atoms, and a benzene ring called an indole, which is attached to a nitrogen atom. Certain indole alkaloids can be toxic to living organisms, while others have medicinal effects. Serotonin and tryptamine, both found in the brain, are derivatives of indole alkaloids. Psilocybin is also an indole alkaloid derivative.

There are three main types of beta-carbolines that are found in the bark of the caapi vine. These are harmine, harmaline, and tetrahydroharmine. Harmine and harmaline are selective and reversible monoamine oxidase inhibitors (MAOIs), meaning they inhibit the metabolism of compounds such as serotonin and norepinephrine. Monoamine oxidase (MO) is an enzyme that breaks down certain neurotransmitters in the brain, such as serotonin, dopamine, epinephrine, and norepinephrine. Harmine and harmaline prevent MO from degrading these amine neurotransmitters. Similarly, these beta-carbolines prevent MO from breaking down DMT, allowing it to remain active in the brain. Studies have shown that beta-carbolines on their own can have slight psychological and physiologic effects, including ataxia, hallucinations, confusion, and agitation. Without these alkaloids, DMT would be inactivated within the liver and gut. It is the combination of the DMT and MAOIs that cause ayahuasca's potent effects.

Users are reported to experience a variety of psychological effects, such as increased self-confidence and new perspectives on conflicts of the mind. Other mental and emotional effects include feelings of euphoria, a loss of touch with reality, near-death experiences, awareness of the sense of self, anxiety, panic, fear, feelings of connection and unity, a peaceful state of mind, and being aware of one's own emotional state. Users also report seeing and contacting higher-dimensional beings and traveling through alternative realities.

Research on the progression of the substance suggests that there is a transcendental cycle of three phases experienced by users. This cycle begins with a change in perception, feelings of vulnerability, and shaking. Thereafter, users start feeling confused, paranoid, and anxious. At this stage, users also start experiencing flashbacks of traumatic memories. The reduction in psychological effects of the chemicals in ayahuasca appears to assist the user in gaining insight into these memories and other personal matters. From there, this fearful stage peaks with intense vomiting that leads to a state of psychological expansion. The next stage is one of transcendence into a feeling of peace, oneness with the universe, and new perspectives on life after death. Users report an alteration in the sense of time and space. The final stage comes with a fading of these visuals, as well as a feeling of being physically exhausted.

Electroencephalograms (EEG) show that patterns of brain activity prompted by the ingestion of ayahuasca share similarities with synthetic serotonergic and dopaminergic drugs. Studies on the movement of blood in the brain show that ayahuasca causes the increased blood flow in the anterior cingulate and frontomedial cortex of the right brain hemisphere. This is the side of the brain that manages and regulates emotion, somatic awareness, and subjective feelings. It also affects the amygdala and parts of the hippocampus, which is believed to explain a user's introspective experience. Hamill et al. (2019) explain that ayahuasca affects a combination of brain regions, causing its unique ability to allow the user to gain new perspectives on previous traumas and negative memories. The effects in each region explain ayahuasca's different aspects:

- **Amygdala**: This prompts the return of old memories, coupled with full emotional intensity.
- **Insula**: This describes the user's heightened states of consciousness, as well as the increase in understanding of self, especially in relation to others.
- **Neocortex**: This explains alterations in perception, behavior, cognition, and reasoning.
- **Default mode network (DMN)**: A decrease in this system may explain alterations in self-awareness and being more aware of one's own thoughts.

As with other psychedelics, ayahuasca appears to disrupt or deregulate normal brain functioning by disengaging associated cortices, allowing users to reform more constructive and positive thought processes. Typical brain networks diverge from one another and brain function becomes increasingly entropic, which means the level of disorder and randomness becomes more chaotic. Highly ordered and predictable brain function and low entropy, suggest less flexible brain activity. High levels of brain entropy point to networks that are more adaptable to new information. It is the prompting of high entropy levels in the brain that explains the psychedelic experiences that occur on ayahuasca, as well as its ability to treat depression.

Therapeutic Applications of Ayahuasca and DMT

In recent years, the popularity of ayahuasca among Western users in North America and Europe has soared. Tourists are inspired to visit South American countries, like Peru, Columbia, and Ecuador to take part in spiritual healing ceremonies led by traditional shamans, a practice sometimes referred to as spiritual or drug tourism.

Users need to keep in mind that, while the plants used to brew ayahuasca are not illegal in the US, DMT is classified as a Schedule 1 Drug under the US Controlled Substances Act. These are drugs deemed by the DEA to have no medicinal value. This means that DMT has not yet been approved for medical use. However, in countries where ayahuasca is used for traditional, spiritual, and religious ceremonies, such as Brazil, its use is legally protected.

Several clinical studies have shown that ayahuasca can positively contribute to the treatment of addiction to substances like alcohol, cocaine, barbiturates, amphetamines, and sedatives. Thomas et al. (2013) showed that a combination of psychotherapy and ayahuasca provided a significant improvement in quality of life, empowerment, and mindfulness in patients suffering from severe behavioral and psychological issues related to addiction and substance abuse. These improvements were reported to be present six months after the treatment, with the majority of participants reporting a decrease in the use of addictive substances such as cocaine, tobacco, and alcohol.

Ayahuasca has also been highlighted as a potential treatment for PTSD, depression, and anxiety, especially with patients suffering from treatment-resistant and recurrent depression. Because of these studies, various synthesized versions of the active compounds are being investigated in clinical trials as potential drug-assisted therapy for depression.

A study by Uthaug (2018) showed that ayahuasca had a significantly positive effect on cognitive thinking styles, extending to more than four weeks after the ceremony. Participants showed a significant decrease in depression and stress, as well as an increase in mindfulness and satisfaction with life. This suggests that just a single dose of ayahuasca has the potential to reduce symptoms of depression in sufferers. While the mechanisms behind the long-term positive effects of ayahuasca are still uncertain, it is thought that the active chemical compounds in the brew promote neuroplasticity and neurogenesis (formation of new neurons in the brain).

The effects of ayahuasca can differ based on several elements, including the strength and composition of the brew, added components, the size and weight of the user, the mental health of the user, the volume consumed, the surrounding environment, and presence of other users, and whether the tea is consumed with other substances.

Clinical Studies and Research

A 2019 study, conducted by Timmermann et al., administered different doses of DMT to a sample group of thirteen participants. None of the participants had previous experience with psychedelic or hallucinogenic substances and had no history of diagnosed psychiatric disorders. Doses of 7 mg, 14 mg, 18 mg, and 20 mg were used, specifically because there is restricted data on the safe intravenous concentration of DMT for humans. All participants also received a saline solution as a placebo at a separate session.

Brain activity for both the treatment and control sessions was monitored with an EEG. Results of the EEG showed a marked decrease in alpha and beta brain waves, as well as an increase in brain entropy in participants who received the treatment. Theta and delta waves appeared when the participants reached the peak of the experience. Different brain waves signify specific activities (Timmermann et al., 2019):

- **Delta**: associated with deep sleep and unconsciousness
- **Theta**: associated with deep relaxation, light sleep, and

meditation; also present during REM sleep and dreaming, and 'visionary' states.
- **Alpha**: linked to daydreaming, meditation, and wakeful relaxation
- **Beta**: linked to wakeful states, concentration, focus, alertness, and engaging in daily tasks

In general, DMT appears to decrease alpha and beta oscillations, while increasing delta and theta oscillations. This is slightly different from other psychedelics, where users experience an overall decrease in brainwaves. Timmermann (2019) suggests that the characteristic breakthrough experience related to DMT is associated with the emergence of theta/delta rhythms and the collapse of alpha/beta rhythms. Here, the team proposes that the brain is prompted by DMT to switch from a wakeful state of processing incoming information to a dreaming state, where the brain starts processing internal information. These changes in brainwaves suggest that the user is experiencing a vivid, completely immersive form of daydreaming when reaching the peak of the DMT trip.

People suffering from depression often show reduced delta oscillations and increased alpha oscillations. Timmermann (2019) believes that the ability of DMT to disrupt these patterns indicates the potential for remodeling brain activity and treating psychiatric disorders.

It has been suggested that, unlike conventional treatments, the effect of psychedelics on the reduction of symptoms of psychiatric disorders is largely influenced by characteristics of setting and set. Setting factors include the surrounding environmental conditions and social context, while the set is a group of subjective factors, such as expectation, personality, intention, and preparation. A recent study by Cruz et al. (2023), aimed to analyze the subjective experiences of participants both with treatment-resistant depression and without, who had no prior experience with ayahuasca. The team wanted to observe the experience of participants without the bias of previous knowledge. Previous studies have shown significant improvements in depression symptoms; however, researchers are concerned that previous ayahuasca experiences may influence the participants' reactions to the treatment.

The randomized, double-blind, placebo-controlled clinical trial assessed 29 participants who had not experienced an ayahuasca session. The results of the trial suggest that the effects of ayahuasca are similar among experienced users (based on other research), as well as naive users. The effects were grouped into five clusters, namely altered states of consciousness, cognitive changes, somatic alterations, auditory experiences, and visual perceptual content. Cruz's (2023) study highlights the common experiences users can expect during an ayahuasca session, regardless of whether it has been done before or not. Interestingly, participants with depression exhibited a significantly higher proportion of negative somatic alterations, indicated by the increased use of words such as dizziness, vomit, pain, and stomach, when compared with those without depression. These results are important to better inform, regulate, and guide the use of ayahuasca in medical settings when treating various psychiatric disorders.

Practical Considerations in Therapy

There have been few reports, thus far, that suggest ayahuasca creates dependence or generates abuse or tolerance. Certain studies note that physical tolerance may be encouraged by regular use, but the effect is not harmful. Interestingly, research on the long-term use of ayahuasca suggests that regular use over an extended period has a lower detrimental effect on cognition when compared with occasional use. These results infer that long-term use of ayahuasca may be neuromodulatory, decreasing memory impairment and incapacitation with administration.

There are several characteristics of ayahuasca that may be responsible for its anti-addictive properties, as highlighted by Liester & Prickett (2012):

- Because ayahuasca reduces dopamine activity in the brain, it reduces the reward-response that usually occurs with an addictive substance.
- The transcendental experiences facilitated by ayahuasca allow the user to re-evaluate their lifestyle, affording them the opportunity to cease detrimental behavior.
- The reduction of dopamine in the brain prevents the synaptic plasticity that is responsible for developing neural circuits related to addictive behaviors.
- Introspection triggered by the experience allows the user to address and heal past trauma, improving their ability to refrain from addictive behaviors.

Because of the extreme vomiting and diarrhea that occurs with its use, there is little risk of toxicity or overdosing with the consumption of ayahuasca.

As with other psychedelics, caution should be exercised when combining ayahuasca with antidepressants and other psychiatric medications, especially SSRIs. A combination of any substance that contains beta-carbolines with SSRIs can cause serotonin syndrome, as described in previous chapters. Research also suggests that taking ayahuasca with ginseng, St. John's wort, or MDMA can cause several harmful effects, like increased blood pressure, increased heart rate, amnesia, coma, collapse, and respiratory distress. Those taking medication for Parkinson's disease and weight loss should also refrain from using ayahuasca.

Research has also highlighted the possibility that ayahuasca can potentially worsen or even trigger psychotic disorders such as schizophrenia. Those with heart conditions and psychiatric disorders should use ayahuasca with caution. If used to treat psychiatric conditions, treatment should be done in a carefully monitored environment and in conjunction with psychotherapy that is facilitated by trained medical professionals.

IBOGAINE

Iboga (*tabernanthe iboga*) is a rainforest shrub native to West and Central Africa, typically growing in the understory of tropical forests. Although it is indigenous to countries like Gabon, the Democratic Republic of Congo, and the Republic of Congo, the plant is now widely cultivated across Central and West Africa. For centuries, iboga has been used in spiritual and religious ceremonies by locals, specifically for tribal initiation rites. Rituals often take place at night, and participants are guided by shamans through the session, which is often accompanied by drumming. Chewing the bark and roots of the iboga plant can produce a hallucinogenic high with dissociative and out-of-body effects.

Tribes like the Fang, Mitsogo, and Punu in the Congo Basin use iboga for its psychoactive properties as part of Bwiti religious ceremonies. The Bwiti religion is a spiritual discipline or study of life that uses iboga as a means of inducing personal development, self-discovery, physical healing, and communication with spirits through healing ceremonies and spiritual rituals (*Bwiti tradition*, n.d., para 1). It is believed that ingestion of the plant allows participants to connect and communicate with their ancestors. Revered for its psychoactive properties, iboga also has social importance, as it is commonly used as an aphrodisiac, a stimulant, and to encourage hallucinations. It is also used to combat thirst and hunger. These days, people also take iboga for treating illnesses such as high blood pressure, HIV/AIDS, fever, and influenza, and as a general tonic.

The plant was first described in Western medicine during the late 1880s by French botanist, Henri Ernest Baillon. French and Belgian explorers had previously recorded its stimulant and aphrodisiac effects when the root is ingested. In the early 1900s, the alkaloid responsible for the plant's hallucinogenic properties, ibogaine, was isolated by the French scientist, Édouard Landrin. Ibogaine was used as a stimulant to promote wakefulness and was believed to hold off fatigue and the need to sleep. Users reported its effects as similar to that of alcohol without the debilitating effects on reasoning. Early research into the effects of ibogaine promoted its effectiveness in stimulating appetite, regulating heart rate, and treating symptoms of fatigue, such as headache, irritability, anxiety, and depression. However, throughout the 19th and 20th Centuries, the misuse of ibogaine increased with a reduction in scientific interest, and it was eventually made illegal in several countries.

Concerns over the abuse and misuse of the substance were raised once again in the early 1960s when reports were made over the potential for ibogaine to cause dependency and endanger human life (Luz & Mash, 2021). Regardless, anecdotal evidence was, at the same time, suggesting that ibogaine was an effective treatment for opioid abuse.

In 1985, ibogaine was used to reduce addiction to substances like morphine, heroin, alcohol, cocaine, nicotine, and amphetamines. A patent was issued for its use to interrupt the addiction cycle and dependency on these substances. Further clinical trials stated that ibogaine therapy was responsible for disrupting dependency and addiction symptoms for up to six months thereafter. Nonetheless, its indiscriminate activity on the brain means that it has several other undesirable side effects, which has led experts to refer to it as a dirty drug.

Chemistry and Pharmacology

The chemical formula of ibogaine is $C_{20}H_{26}N_2O$. It can also be extracted and semi-synthesized from other alkaloids such as voacangine. It is an indole alkaloid, a group of plant-based compounds that have a bicyclic structure, which is made up of a pyrrole ring that is fused to a benzene ring. Ibogaine is rapidly metabolized primarily in the liver and the gut wall by the cytochrome enzyme CYP2D6 into noribogaine (12-hydroxyibogamine), its active metabolite. Metabolization peaks at four hours after administration of the substance, when noribogaine concentrations are at their highest. Noribogaine is then eliminated within 24 and 30 hours after administration. Because it is a lipophilic substance, ibogaine accumulates in the body's fat.

The alkaloid responsible for its stimulant and excitatory effects is a crystalline alkaloid that has similar effects on the brain as SSRIs such as fluoxetine and other antidepressants. Although the specific action of ibogaine in the brain is still largely misunderstood, it is believed to act on several reward-aversion regions of the brain, namely the amygdala, habenula, nucleus accumbens, and raphe nucleus. It works by deregulating receptor signaling in the neurotransmitter systems in these brain centers, which triggers its notable anti-addictive effects.

Ibogaine is thought to have a moderate to weak affinity for several neurotransmitter systems in the brain, similar to other psychedelic substances like MDMA. It blocks NMDA receptors the same way that ketamine does but also activates the serotonin 2A receptor, like LSD, psilocybin, and DMT.

It works primarily by binding to a broad range of receptors and acts as both an agonist and antagonist. This gives it its hallucinogenic and anti-addictive characteristics. Ibogaine targets NMDA (N-methyl-D-aspartate), sigma (σ), mu (μ), kappa (κ), nicotinic, and acetylcholine receptors, as well as serotonin and dopamine transporters. It is these antagonistic effects on various receptors that cause ibogaine to modulate and reduce these neurotransmitters.

Its propensity to catalyze several reactions and act on different neurotransmitters, called enzyme promiscuity, has prompted researchers to isolate the molecule responsible for its anti-addictive properties. By isolating the specific molecule that gives it this ability, there is a reduction in the potential negative side effects, such as cardiac toxicity, that are experienced with ibogaine as a result of its enzyme promiscuity.

Therapeutic Applications of Ibogaine

Ibogaine was classified as a Schedule 1 substance in 1970 by the US Food and Drug Administration (FDA), largely due to its hallucinogenic effects and its appearance as an illicit drug. Its classification means it is prohibited from use in a non-research setting, has no recognized medical benefits, and has a high risk of abuse.

Nonetheless, ibogaine's ability to treat opioid dependency is boosting its popularity. Currently, there are few, if no, successful and long-term treatments for prescribed and illegal opioids, despite what experts are calling the opioid crisis in the US. Recent figures suggest that the US is experiencing over 130 lethal opioid overdoses every day.

One of the characteristics of ibogaine that has promoted its popularity as a treatment for psychiatric disorders is that it is believed to eliminate withdrawal symptoms and substance addiction for extended durations after a single dose. Although this evidence is largely anecdotal, there have been several animal studies that have supported this theory. Ibogaine is under research as an alternative to methadone for treating heroin dependency. Currently, methadone is the more common treatment for opioid dependency; however, because it is a long-acting opiate, patients often experience withdrawal and detox symptoms when weaning off it.

In countries where the use of ibogaine is not prohibited or regulated, such as New Zealand, Mexico, Costa Rica, and the Netherlands, numerous clinics have opened to those requiring treatment for substance abuse, addiction, and dependency. Ibogaine treatment clinics, many of them makeshift and informal, are springing up to treat many of those suffering from the impact of the global opioid crisis.

Clinical Studies and Research

Because of its non-selective activity on receptors in the brain, researchers are working on isolating the properties of ibogaine that can be used safely to treat addiction, without the numerous undesirable side effects. A team of scientists has used a process called docking to isolate the compounds that work on the serotonin transporter (SERT) specifically. Docking involves virtual testing of various chemical structures to identify similar molecules without having to synthesize them in the lab. After several docking experiments, the team managed to identify two main molecules that were even more effective than ibogaine in treating addiction, depression, and anxiety. A drug synthesized from one of these molecules appeared to act as a SERT similar to ibogaine but without the psychedelic and other detrimental side effects. Instead of acting indiscriminately throughout the brain's neurotransmitter system, the isolated drug was able to selectively impact targeted receptors and transporters.

Various studies conducted in Mexico and New Zealand showed that participants struggling with opioid dependency who received a single dose of ibogaine were still in remission after one year post-treatment (75% and 50%, respectively) (Barsuglia et al, 2018). A recent clinical trial (Cherian et al., 2024) was conducted on 30 participants with mild traumatic brain injury (TBI), which resulted in several psychiatric brain disorders such as PTSD, depression, and anxiety. Quality of life was negatively impacted by the resulting psychiatric disorders, and participants were experiencing significantly high levels of disability. Ibogaine was administered orally to the participants, in conjunction with intravenous magnesium, over a period of two hours. Treatment was done in the presence of medical staff. Results showed a significant improvement in PTSD (88%), depression (87%), and anxiety (81%), both immediately post-treatment, as well as a month after treatment (Cherian et al., 2024). Participants also experienced an improvement in cognitive processing abilities, like memory and concentration. Additionally, no detectable incidents of adverse side effects, such as extreme changes in heart rate or rhythm, or unstable blood pressure, were reported. However, some of the participants recorded mild side effects such as anxiety, insomnia, nausea, headache, and an increase in blood pressure.

A Brazilian study was done on 75 participants in 2014, all of whom suffered from various addictions to substances such as alcohol, cocaine, crack, opioids, and cannabis. Approximately 54 of the participants were reported as poly-drug users. All participants were encouraged to take part in a drug-assisted therapy treatment, whereby they received doses of ibogaine that were augmented with psychotherapy. A retrospective analysis of the participants, six months after treatment, showed that more than 60% of participants were abstinent from all substances (Schenberg et al., 2014). While the study showed that ibogaine can be effective in reducing or removing substance-related cravings, as well as decreasing opioid withdrawal symptoms, the effects are only short-lived and only effective in large doses. Large doses, unfortunately, can cause severe psychedelic states, as well as several other physiological issues, such as paralysis, a slowing down of the heart rate, seizures, and even death.

A review was done on the adverse effects of ibogaine, based on the analysis of eighteen studies conducted between 2015 and 2020. The results showed that most of the acute effects caused by the administration of ibogaine, in varying doses (known and unknown) were cardiac (specifically, irregular heart rhythms, called QTc prolongation). Other acute effects included gastrointestinal and neurological issues, most of which lasted less than 24 hours. However, persistent cardiac alterations, as well as other psychiatric and neurological issues prevailed for longer than 24 hours (Ona et al., 2022).

A 2019 study done on the effect of ibogaine on the brains of rats (Marton et al., 2019) highlighted another characteristic of the substance that may explain its anti-addictive properties. Marton et al. (2019) showed that ibogaine worked on a mesencephalic region of the brain called the Ventral Tegmental Area (VTA). The VTA hosts one of the main reward pathways in the brain, modulating reinforcement, addiction, and motivation. Within the VTA, a high-density bundle of dopamine-releasing neurons sends information to the nucleus accumbens. If the VTA encourages the release of dopamine into other regions of the brain, the experience is one of pleasure, which then reinforces behavior that is associated with the reward. The research team administered ibogaine to rats and analyzed the resultant effects on the expression of neurotrophic factors in the brain, such as glial cell-derived neurotrophic factor (GDNF).

GDNF is a protein that promotes and modulates the survival of neuronal cells, specifically dopaminergic neurons. Ibogaine was observed to stimulate the production of GDNF under certain doses. By increasing GDNF, the survival, and function of dopaminergic neurons is encouraged. In people who have addiction disorders, these neurons may have been damaged. GDNF can assist in mitigating this damage and promoting the growth of new neurons. Also, by restoring correct functioning in the dopaminergic neurons, the craving and withdrawal symptoms associated with drugs can be decreased. It is this action that gives ibogaine its anti-addictive properties. Finally, the enhanced functioning of dopaminergic neurons can facilitate the restoration of healthy reward processing within those regions of the brain, decreasing the reinforcement of drugs. Marton's (2019) research showed dosing rats with ibogaine simultaneously modified the expression of GDNF, as well as other neurotrophic factors, some of the effects lasting after the 24-hour treatment period.

Practical Considerations in Therapy

Currently, the use of ibogaine is not approved in the US. Its unregulated use across the world for the treatment of opioid addiction has led to several deaths. Ibogaine can place stress on both the kidneys and the liver. However, its stimulant effect is what puts such strain on the heart, causing the heart rate to speed up to a point where the user goes into cardiac arrest. High doses of ibogaine can also cause seizures, irregular heart rhythms, decreased respiratory function, and a loss of consciousness.

Research shows that a total of 33 deaths linked to ibogaine use have been reported, most occurring 24 hours after ingestion of the substance (Luz & Mash, 2021). However, it is notable that these fatalities occurred in settings deemed unsafe or unregulated, where patients did not have access to medical equipment or monitoring devices. In addition, it is likely that patients ingested ibogaine while using other substances like alcohol or drugs. Patients were not checked beforehand for pre-existing cardiac conditions and other predisposing comorbidities. It is also possible that the ibogaine administered was not a pure form of the substance.

Ibogaine is typically used as a once-off treatment for addiction, as well as to reduce symptoms of withdrawal; however, this is often done without supplemental psychotherapy, which means patients are forced to return for more treatment after relapsing. As with other psychedelic substances, such as ayahuasca and MDMA, it is highly advised that treatment is done in conjunction with psychotherapy. The therapy can assist patients to address the underlying issues behind their addiction disorders.

Unfortunately, the surge in interest in ibogaine has caused unchecked poaching of the iboga plant in Central Africa, threatening both the culture and history of the communities that use it, as well as the state of the environment. Researchers are hopeful that further studies into the development of synthetically similar molecules can prevent the indiscriminate plundering of the natural sources of ibogaine.

Currently, the FDA has halted the approval of ibogaine in clinical trials, mainly over the safety concerns of the substance. However, ibogaine and its synthetic compounds show great promise in the treatment of addiction. Experts suggest that if ibogaine is administered under professional and qualified medical supervision, the benefit-risk ratio is favorable for its use in opioid withdrawal management (Luz & Mash, 2021). Other research also suggests that administering ibogaine with magnesium may reduce the likelihood of fatal cardiac arrhythmia (Cherian et al., 2024).

It is useful to note that the effects of ibogaine have three stages, namely the acute phase, which occurs 4-8 hours after ingestion; the evaluative phase, occurring 8-20 hours later; and the residual phase, which occurs 24-72 hours after ingestion. During the acute stage, participants are expected to experience a state of dream-like consciousness. Thereafter, participants are encouraged to evaluate and contemplate their experiences during the previous stage. The final stage requires participants to assimilate their surrounding environment with the lingering effects of phases one and two. This is important information for future use in a controlled environment and to ensure safety precautions.

Administration of ibogaine causes the following symptoms, which can vary in severity based on the dosage:

- hallucinations
- anxiety
- low blood pressure
- seizures
- paralysis
- irregular heart rate
- cognitive disruptions
- difficulty breathing

In general, ibogaine has the potential to successfully reduce the craving for opioids and mitigate withdrawal symptoms. While researchers suggest that ibogaine can be used to treat addiction, it must be kept in mind that it is not a cure; rather, it is a mechanism to interrupt addiction. However, there are several concerns regarding its use outside of medical supervision. As it stands, ibogaine can have severe and negative side effects if the dosage administered is too high. An appropriate dose is highly dependent on the age, health, and other characteristics of the user. If used responsibly and with medical assistance, ibogaine can be a successful, safe, and transformative treatment for opioid dependence. For it to be used safely and effectively, treatment needs to take place in controlled environments with medical professionals. Controlled clinical studies need to be conducted to provide strict guidelines and regulations for its use.

MESCALINE

Mescaline is the naturally occurring active ingredient found in a small spineless cactus called the peyote cactus (Lophophora williamsii). Peyote has been used by Native Americans for centuries as a prominent element of their spiritual and cultural practices. Evidence of the use of peyote has been found going back to more than 5,700 years ago. As such, peyote is often considered one of the oldest recorded examples of a hallucinogen.

The common name, peyote, is derived from the Nahuatl language, which is in the Aztecan language family. The plant is still considered sacred in more than 40 Native American tribes across Canada and North America. While rituals and ceremonies differ slightly between tribes, the ingestion of peyote is often accompanied by cleansing rituals, meditation, and chanting. Currently, peyote is classified as a Schedule 1 controlled substance by the Drug Enforcement Administration (DEA), which means it has no legally recognizable medical use in the US, has no safety guidelines for use under medical supervision, and has a high potential for misuse. However, the Native American Church is legally authorized to use the plant for religious ceremonies, as well as for treating physical and psychological afflictions.

The alkaloid compounds of peyote were considered useful in medical applications in France in the early 1920s. It was first synthesized by Austrian chemist, Ernst Späth. Thereafter, the compound was isolated in the 1930s when it was studied by the French neurologist, Henri Ey. It was at this time that mescaline was being used in experimental psychiatric therapy treatment. However, the possession of peyote was made illegal in many states in the US in 1930. It was banned countrywide by the federal government in 1967.

Chemistry and Pharmacology

In addition to the peyote cactus, mescaline is also found naturally in small quantities in various plants, including the guajillo shrub (Senegalia berlandieri), the cactus Gymnocalycium chubutense, San Pedro (Echinopsis pachanoi), and Peruvian torch (Echinopsis peruviana), among others. However, it is predominantly isolated for use from the peyote cactus. When dried, the cactus contains approximately 3-6% mescaline. It is the psychoactive characteristics of mescaline that give peyote its hallucinogenic effects. Anecdotal data suggests a form of synesthesia occurs with the ingestion of peyote, where users report a mix of senses. Senses are enhanced and time can become distorted.

Interestingly, reports suggest that peyote be consumed within a controlled and ceremonial environment to reap its benefits. In 1959, Anthony Wallace, an American-Canadian anthropologist, researched the effects of peyote in a cross-cultural setting. He noted that those from indigenous groups with a history of peyote consumption largely benefited from these experiences with positive spiritual experiences and encounters. However, when consumed by Western or European-descended people, either in an uncontrolled or recreational setting, Wallace observed that users experienced negative and gloomy hallucinations, fear, and distress. Some of these uncomfortable experiences stayed with the users for years. This led Wallace to advocate for tighter control of the substance, especially in clinical trials, encouraging these trials to consider the cultural aspect of the substance when used. These early observations paved the way for a better understanding of psychedelics as a whole. Researchers now almost universally accept that the mindset of the user and their environment plays a significant role in determining the outcome of their experience with psychedelics.

Peyote is native to southern Texas and northern Mexico in North America. It is a perennial succulent that grows in limestone desert soils. Peyote buttons, which are dried sections of the cactus, are used in several Native American religious ceremonies to induce color hallucinations. These buttons are formed by the round wart-like projections or nodules of the cactus, each of which holds between 40 to 50 mg of mescaline. Unfortunately, due to the increase in leisurely use of the plant for its psychedelic effects, Texan populations have almost been decimated by collectors. As a result, to possess or carry peyote specimens, a federal permit is required.

Chemically, mescaline is a form of amphetamine, specifically a ring-substituted amphetamine. This means it is based on the amphetamine structure with one or more hydrogen atoms replaced or substituted to form a similar, but differently-formed compound. Its chemical name is 3,4,5-trimethoxyphenethylamine and its molecular formula is C11H17NO3. Mescaline is an active alkaloid form of phenylethylamine. Phenylethylamine is an organic compound that is also produced in small amounts in the brain as a central nervous system stimulant. It acts as a neuromodulator, triggering the release of certain neurotransmitters, including serotonin, dopamine, and epinephrine. Because mescaline has a substitution on the benzene ring, it has greater hallucinogenic properties, rather than the stimulant effects of amphetamines.

Mescaline can be synthetically produced as mescaline sulfate. This is a white crystalline powder that is often administered as an infusion, also called peyote tea. However, mescaline can also be smoked or insufflated (a vapor or powder that is blown into the nose or mouth).

Because the majority of the chemical is distributed to the kidneys and liver, the concentration is delayed in the blood. This is what causes its half-life to be increased and its effects to be delayed. It is not readily absorbed across the blood-brain barrier, which is why higher doses are required to reach similar effects as other psychedelics.

Generally, a dose of 5 mg of synthetic mescaline per 2.3 pounds of body weight induces hallucinations. This is equivalent to four to twelve peyote buttons. Mescaline is rapidly absorbed in the gastrointestinal tract and effects start showing approximately 30 minutes after ingestion. Thereafter, psychotic effects occur after approximately three to four hours. Its plasma half-life is six hours, but hallucinogenic effects can last for up to eight hours. Almost 90% of the dose is usually excreted in the urine after 24 hours (Dasgupta, 2019).

Mescaline has a predominant influence on the right hemisphere of the brain. A full dose is considered 200-400 mg of mescaline. At larger doses (300-500 mg), mescaline can induce auditory and visual hallucinations, anxiety, depression, and depersonalization (feelings of disconnection from one's emotions, body, and the surrounding environment). Physical symptoms of ingestion include tremors, sweating, and nausea. In general, both physical and psychological effects closely resemble those produced by the ingestion of LSD (see Chapter 3). The rapid breakdown of mescaline in the body means its stimulant effects can be weak and short-acting.

Therapeutic Applications of Mescaline

Much like other psychedelics, mescaline is considered a 5HT2A/2C receptor agonist. This means it impacts the brain by working on serotonin receptors by binding to them and stimulating them. The clinical effects of mescaline are therefore very similar to that exerted by LSD and psilocybin. These include hallucinations, psychoses, depersonalization, and euphoria. However, unlike these similar substances, its psychoactive properties remain largely misunderstood.

Because it activates serotonin receptors, mescaline can cause alterations in consciousness, mood, emotions, and perception. As with other psychedelics, it can cause both closed- and open-eye hallucinations or visual phenomena, an altered perception of time, and changes in self awareness and thinking processes. Unique to the substance is its ability to create what is called the geometrization of three-dimensional objects, which means that objects are assigned geometric patterns and can appear flattened and distorted. Users also report feeling a connection to a higher being or the ability to communicate with deities and other worlds. Mescaline also lightly activates dopamine and noradrenaline receptors, causing slight anti-anxiety effects that encourage prosocial behavior and increased movement and mobility.

Mescaline can promote feelings of well-being, which means it has been considered for use in treating various psychiatric disorders such as anxiety and depression, specifically in clinical settings. While numerous studies over the years have suggested using psychedelics to treat alcoholism, OCD, and PTSD, research on mescaline is limited.

Although little work has been done on mescaline in substance-assisted psychotherapy, peyote is used by several Native American tribes to cure ailments such as burns, wounds, rheumatism, and snakebites. Its antibiotic properties have been attributed to the presence of hordenine, which is another alkaloid present in the plant. Hordenine is found primarily in the roots of the peyote cactus. Another alkaloid, pellotine, can have sedative or hypnotic effects, while the anhalinine alkaloid is a stimulant. It is this combination of various alkaloids that is believed to enhance the psychiatric effects of mescaline.

Mescaline doesn't appear to cause cytotoxicity, which makes it safe for human consumption. However, several cases of intoxication have been reported, whereby patients are sedated and given supportive measures as treatment. Nonetheless, these reports are anecdotal, which means it is not certain whether mescaline was taken on its own or with other substances.

Clinical Studies and Research

Early studies on mescaline showed that users can develop a slight tolerance to mescaline; however, this tolerance decreases after three to four days of no use. Sensitivity to the substance is thus restored thereafter.

At doses over 200 mg, users of mescaline can exhibit nausea and vomiting, although this has often been attributed to the bitter taste of peyote. At doses over 300 mg, users experience hypersensitivity to noise, a loss of body orientation, and tingling in the fingertips.

A study conducted in 1955 on a group of patients diagnosed with schizophrenia aimed to analyze the effect of 500 mg of synthetic mescaline on the disorder. However, only a single participant showed any significant improvements in their symptoms lasting a year post-treatment, while only three participants experienced temporary relief from symptoms. The remaining participants, in fact, experienced a worsening of their symptoms or reactivation of psychotic episodes. A follow-up on the study on a further set of patients with schizophrenia showed that only one patient received enough respite from symptoms to be discharged from the trials, while the majority of the remaining participants exhibited no changes in their symptoms.

Several decades later, a paper was published that suggested mescaline has a positive impact on alcoholism by assisting patients to overcome the disorder. The substance was given to participants in the study in conjunction with ritual ceremonies and cultural talk therapy. Results showed that many of the participants were able to overcome their addiction after just a single dose. A follow-up on this study more than twenty years later showed that participants of the 1974 study who had used peyote to overcome their alcoholism had significantly higher levels of well-being and lower scores for anxiety, depression, and psychological distress. This was in comparison with other participants who had not used peyote or engaged in cultural therapy, even those who were lifetime abstainers from alcohol.

In the late 1980s, clinical studies were done on the effect of mescaline on the brain. Brain imagery showed that 500 mg of the substance created activity in the limbic system of the right brain hemisphere that was similar to that caused by schizophrenia. Later, a second study reported that 500 mg of mescaline caused an acute psychotic state in participants around four hours after ingestion. Also, left hemisphere and right visual field functions improved, while left visual and right hemisphere functions deteriorated.

A review of anecdotal reports obtained from over 450 people that was conducted in 2021 wanted to analyze the effect of mescaline on the symptoms of mental illness. An anonymous online survey requested information on the use of mescaline in a naturalistic setting, especially in its benefits to mental health. These self-reported surveys assessed improvements in mental health disorders such as depression, anxiety, PTSD, alcoholism, and drug use. Those participants who reported historic incidences of mental health issues also reported a subjectively significant improvement in their symptoms (86%) (Agin-Liebes et al. 2021). In addition, these participants also reported higher incidences of ego dissolution and psychological insights; those psychological insights appeared to correspond with improvements in anxiety, depression, and substance abuse.

In 2022, a clinical trial was started in Switzerland that looked to analyze the effects of moderate to high doses of mescaline on the brain in comparison with other psychedelics, namely LSD and psilocybin. The study aimed to better understand how these psychedelics affect the stimulation of the 5-HT2A serotonin receptors, with a focus on using them to treat various psychiatric conditions. Various brain-imagery techniques such as magnetic resonance were used to investigate the physiological and neuronal effects of the substances, while psychometry was used to analyze psychological variables. The research was conducted as a double-blind, placebo-controlled, randomized study with four treatment conditions. Several variables were measured throughout the trial, including subjective states of consciousness, blood pressure, heart rate, levels of oxytocin, and mood rating, among others. A total of 32 participants completed the trial.

Results showed that mescaline had the longest duration effect of approximately eleven hours, in comparison with LSD, with eight hours, and psilocybin, with five hours. In general, the peak effects of mescaline also took longer to be reached, in comparison with LSD and psilocybin. The researchers concluded that, regardless of their chemical and pharmacological differences, there were no significant differences in the subjective experiences of the participants. This conclusion has potential consequences for the use of various psychedelic substances in the treatment of psychiatric disorders, where those with the least side effects can be considered preferable over others.

There are several pharmacological differences between mescaline, LSD, and psilocybin, which are outlined in the table below, as per Dinis-Oliveira et al. (2019):

	Mescaline	**LSD**	**Psilocybin**
Average dose	20-500 mg	0.05-0.2 mg	20-40 mg
Potency in comparison with mescaline	-	2,000 times stronger	20 times stronger
Time to onset of effects	1-3 hours	30-40 minutes	20-30 minutes
Duration of effects	More than 10-12 hours	8-12 hours	4 hours
Occurrence of flashbacks	Rare	Common	Rare

Practical Considerations in Therapy

Mescaline is approximately 30 times less potent per dose than psilocybin and is the least potent of the naturally occurring psychedelics. While the lower levels of potency may prevent overdose, it can lead to complications in applications in a medical setting, as patients would need to receive higher doses than other psychedelics to have an effect. There are no reports of addiction to or dependence on mescaline; however, healthcare practitioners do display concern that it can have the potential for misuse due to its hallucinogenic properties.

It is considered one of the more favorable psychedelics for the treatment of psychiatric disorders, specifically because of its long-acting duration. For example, the effects of psilocybin last for approximately four to six hours, while mescaline can have an effect for up to eight hours. Clinical studies suggest that slower-acting psychedelics have a longer-lasting therapeutic effect. This means that psychedelics like mescaline may have greater subjective effects on mood and cognition than faster-acting alternatives. Also, because the psychedelic effects take slightly longer to occur after ingestion, it can help patients or users to more readily surrender to and accept the process. This can then lead to a great therapeutic benefit.

If mescaline is taken in extremely high doses, it can result in severe toxicity and even death. Users can also experience what is commonly known as a bad trip, where hallucinations can be overwhelmingly frightening and involve negative emotions. Because of the distortion of time that can occur simultaneously, users report feeling as though they are trapped within these emotions and visions. Other reported physical side effects of taking mescaline include:

- shivering
- fever
- chills
- tremors
- increased blood pressure
- increased heart rate
- vomiting
- nausea
- dilated pupils

In very rare cases, users report a disorder called persistent psychosis, where they may experience more than six months of symptoms similar to psychotic disorders like schizophrenia. However, long-term studies, such as those discussed above that were conducted with Native Americans with direct consumption patterns, show no evidence of long-term cognitive or psychological problems. In addition, there are no reports of long-term users experiencing flashbacks, or what is known as hallucinogen-persisting perception disorder.

Pharmaceutical companies are looking to synthesize a version of mescaline that reduces the potential for negative physical side effects like vomiting and nausea. Recently, a synthesized version of mescaline has been developed in a California-based laboratory that is being touted as a treatment for alcohol use disorder. Another laboratory based in Vancouver is looking at a synthetic version of mescaline to treat inflammation and a range of psychiatric conditions, with a minimized potential for vomiting and nausea.

SALVINORIN A

Salvinorin A (SA) is the active ingredient of the psychoactive plant, *salvia divinorum.* It is secreted as a resin from trichomes (hairy epidermal cells) on the plant's leaves. Also called the diviner's sage or Mexican mint, *s. divinorum* is a semi-tropical perennial herb that can grow up to 22 feet. It is part of the mint family (*Lamiaceae*) and is native to Mexico, specifically the southwest region of Sierra Mazateca, Oaxaca. Its square-shaped, hollow stem is characteristic of the plant, as are its white and blue flowers. The plant reproduces primarily through vegetative means.

The herb has been used for centuries by Mazatecan people in a medical and cultural capacity and to induce hallucinatory effects. Typically, the plant is rolled into a cigar-like shape and sucked or chewed. In another method, the leaves are ground and dried to be used as a tea infusion. It is also used to treat medical conditions such as headaches, gastrointestinal disorders, and rheumatism. *S. divinorum* is also useful in treating insect bites, eczema, menstrual cramps, and cystitis. Because it has been highlighted as a successful treatment of depression and alcohol addiction, its pharmacological properties are being investigated.

The hallucinogenic effects of the plant were reported by a Spanish researcher in the field in the 1950s, who noted that chewing the plant gave the user a sense of depersonalization, dysphoria (a sense of unease), altered perceptions of sound, light-headedness, visual and auditory imagery, and various tactile sensations. While the field research reports showed no actual hallucinogenic effects, current anecdotal reports suggest that hallucinations are produced at doses of 250-500 micrograms of SA. However, its hallucinogenic effects do not appear as significant as those of the synthetic compound, LSD. In addition, brain imagery techniques show that SA has potentially different psychomimetic effects (hallucinogenic effects similar to psychoses) compared with other hallucinogens like mescaline and ketamine. While it is considered highly potent, it is rapidly degraded in the gastrointestinal tract.

While *s. divinorum* has been used culturally for hundreds of years, it is starting to gain popularity as a recreational drug. It is now administered through inhalation of smoke or vapor. The effects of SA can be seen as little as a minute after administration, with a subsiding of effects after 15 minutes. The effects of SA include dream-like experiences, a deep insight into personal situations, a feeling of the presence of other entities, auditory hallucinations, speaking in tongues, altered visual perceptions, out-of-body experiences, flushing, a loss of muscle control, panic, headaches, and loss of consciousness.

The dried leaves of *s. divinorum* are readily available from Internet suppliers or shops in countries and states where use is not prohibited. Generally, the seeds are almost impossible to obtain and plants are propagated through cuttings.

Chemistry and Pharmacology

Salvia divinorum was first brought to the US in 1962, where it was used primarily for its hallucinogenic effects, rather than for religious and cultural purposes, as in Mexico. The main active component of *S. divinorum*, specifically called salvinorin, was isolated in 1982 by researchers. It was further divided into two compounds, namely salvinorin A and salvinorin B. Salvinorin A is a neoclerodane diterpene compound with psychoactive properties, while salvinorin B differs from A in that it has a slightly different chemical structure and no psychoactive components. In fact, research has shown that salvinorin A is also significantly more potent than salvinorin B. While salvinorin B has none of the hallucinogenic components of A, it is still considered an important substance for its potential as a synthetic analog in a therapeutic setting.

In general, diterpenes have various actions, including anti-inflammatory, antimicrobial, and immunomodulatory effects. Because of this, they are of great interest in treating cardiovascular disorders, as well as arthritic conditions. Generally, diterpenes are composed of four isoprene units and twenty carbon atoms arranged in a bicyclic structure. They are naturally occurring organic compounds. Changes in the side chains and functional groups give various diterpenes their specific biological properties.

The chemical formula of SA is $C_{23}H_{28}O_8$. It cannot form soluble salts, which may limit its use in clinical settings as a formulated drug. The integrity of the isolated compound is also limited, which raises concerns about its suitability for medical use. Further research into its chemical stability is required, especially if it is to have the expected effect on the kappa opioid receptors (KOR), which is specifically required to treat selected disorders.

In addition, an effective mode of administration needs to be developed to ensure its effectiveness; oral ingestion has proven ineffective due to its susceptibility to enzymatic degradation. Intravenous administration also poses an issue, specifically because of SA's poor aqueous solubility. Thus far, only pharmaceutically unacceptable solvents have been used in *in vivo* studies where the substance was administered intravenously. However, the intravenous administration of the substance allows therapeutic doses to be tightly controlled, which is preferable in clinical use. Further research on the development of SA as an intravenous nanocrystal may provide viable potential, especially for use in treating neurological disorders. While administration of the substance through smoking or vaporization shows rapid intake into the brain, accompanied by intense psychotropic effects, it is difficult to curate dosage levels with this process (Orton and Liu, 2014).

Salvinorin A is a potent KOR agonist. It is the only known compound that is so widely available, having a selective high-efficacy agonist effect at KORs. It is also a partial agonist of the dopamine D2 receptor. KORs exert their effects on the brain and on the gastrointestinal tract. They modulate various areas within the central nervous system and regulate several functions, including emotional, perceptual, cognitive, reward, and neuroendocrine. They are also considered to have an effect on disorders such as obesity, depression, and anxiety. It is thought that their activities in the brain can influence and normalize cognitive and behavioral modifications.

KORs modulate and control mood and motivation and are found in corresponding brain regions, including the amygdala, basal ganglia, thalamus, and hypothalamus, among others. If the KOR system is activated, several responses occur, including anxiety, stress, aversion, depression, submissive behavior, sedation, and reward-seeking behavior (with a focus on addictive drugs). In the same way, activating the KOR pathways in the brain encourages and stimulates the addictive qualities of drugs of abuse, such as cocaine and morphine. Research on KOR agonists, which block the activation of the system, shows a decrease in these behavioral responses. By administering synthetic KORs to the brain, dopamine release can be decreased while increasing the activity of dopamine transporters. This, in effect, reduces the concentration of extracellular dopamine. Simply put, KOR agonists directly oppose the effects of addictive substances. In addition, they appear to have very little potential for abuse.

Non-human trials suggest that the effect of KOR agonists occurs during the intoxication or binge stages of the addiction cycle. The binge stage is followed by the withdrawal effect, and then the anticipation of the next binge stage. In essence, KOR agonists appear to decrease the rewarding effects of drugs by modulating the projection of dopamine from the nucleus accumbens to the ventral tegmental area. They also seem to decrease the anxiety related to addiction and the potential for relapse.

In fact, as it stands, SA is the only known non-nitrogenous KOR that occurs naturally. It has no impact on 5-HT2A serotonin receptors, unlike other psychedelics such as psilocybin and LSD. SA works to activate KORs and, in turn, affect the regulation of dopamine levels in the brain. Research on its effects on several receptors, ion channels, and transporters in the brain shows it has an affinity to cannabinoid type 1 receptors in the cannabinoid system. Its hallucinogenic effects on humans are believed to be a result of its activity with KORs. Because it has no effect on serotonin receptors and does not stimulate the release of dopamine, it has little potential for dependence.

Considered the most potent of the hallucinogens, SA has active doses from 200-500 micrograms. Various hallucinogenic effects include altered perceptions of reality, hallucinations, and a sense of time travel. The leaves can be smoked or chewed, allowing the psychoactive chemicals to be absorbed by the mucous membranes in the mouth. Oral ingestion causes intense but short-lived hallucinations, as well as sedation, and dysphoria.

Therapeutic Applications of Salvinorin A

While SA is legally sold across the US (in other words, it is federally lawful), certain states, including Maine, Missouri, Tennessee, and Louisiana, have prohibited its sale, as well as its use and possession thereof. Thus, it must be noted, especially for clinical use, that usage, sale, and possession laws vary from state to state. In certain states, the plant is lawful for any use other than that of the purpose of human consumption. In others, only those over the age of 18 years are allowed to consume the plant. For example, in Delaware, the plant is classified as a Schedule 1 substance, which means it has a high potential for abuse and no recognized medical use.

As of May 2024, the Drug Enforcement Agency (DEA) is in the process of reviewing the federal legality of SA in order to make it a controlled substance. Within the Controlled Substances Act, the substance will then be placed under one of the five schedules, according to its potential for medical use, as well as the potential for misuse and safety of use. Currently, *s. divinorum* is listed as a poisonous plant in the US, while the DEA maintains that there are no approved medical uses for it.

Salvinorin A has a short half-life and a rapid onset of action in humans, reaching peak plasma concentration within ten minutes of administration and returning to baseline levels within half an hour. Users achieve psychoactive effects that are short-lasting, namely, within seconds to minutes of onset. Generally, in both human and animal models, any treatment of SA that lasts longer than 20 minutes has no significant changes in behavior. This has implications for the use of SA in treating addictive behaviors, especially with SA analogs. The aim is then to develop SA analogs that have an improved pharmacokinetic profile, especially with novel compounds with varying opioid binding characteristics.

Two salvinorin analogs are showing promise as effective alternatives to SA. The first one is ethoxymethyl ether salvinorin B (EOM Sal B) and the second one is methoxymethyl ether salvinorin B (MOM Sal B). Both analogs are potent, selective compounds that have shown a greater affinity (*in vivo*) to plasma protein and increased metabolic stability. This means they are likely to have a longer duration of effects. *In vitro* studies show that the activity of these two analogs is up to seven times more potent at the KORs than SA. Researchers suggest that while this may be the first known study that notes the significant similarities between synthetic analogs and SA, it is also thought that they can be used as a substitute for other psychedelics like ketamine and LSD.

Research on the effects of SA has shown its potential in the biomedical field for the treatment of pain, specifically as an analgesic. This is the same for its metabolites, as well as its semi-synthetic metabolites. Even at low doses, the substance seems to reduce the occurrence of pain symptoms in rats, such as escape behavior and licking. However, studies on the effect of SA in treating pain in humans have not been done. Research does suggest that its low potency and very short half-life may present restrictions for its use in effective pain treatment.

Salvinorin A has been shown to reduce brain activities that are related to reward, learning, memory, and tolerance. It is these properties that highlight it as a viable treatment for pain, without the addictive potential prevalent in other analgesics.

Generally, oral administration of SA reduces its effects, specifically because it is highly susceptible to degradation by enzymes in saliva, as well as in the gastrointestinal tract. It is also metabolized very quickly and has a short-binding affinity to receptors. As a result of these restrictions, medical studies are looking at producing new derivatives and synthetic analogs that can produce the required analgesic and anti-inflammatory results but with a longer-lasting duration.

As it stands, there are approximately 600 semi-synthetic SA analogs that have proven effective in treating inflammation and pain. These include herkinorin, which targets *mu*-opioid receptors, PR-38, which targets both *mu* and *kappa* receptors, and beta-tetrahydropyran salvinorin B, which targets *kappa* opioid receptors. These analogs don't have the psychoactive properties of SA and focus only on anti-inflammatories and analgesics. Herkinorin is able to produce its anti-inflammatory and analgesic effects with the desensitization that results in drug tolerance. In addition, beta-tetrahydropyran salvinorin B can be ingested orally, without the fast metabolization of SA. This means it has a much longer-lasting effect than SA, while also regulating pain and other pathophysiological conditions, giving it great potential for clinical use.

It must be noted that consumption of SA may have long-lasting effects of withdrawal with chronic use. These can include anxiety, irritability, and malaise. In addition, various increases in social and interpersonal problems with users can indicate the potential for substance use disorders. Researchers do suggest that because of the limited data available on the effect of repeated administrations of SA in humans, exposure can differ widely across individuals, varying according to prior trauma or stress, existing drug use, or genetic variation.

Clinical Studies and Research

Few studies have been done on the effect of SA on the human brain. Animal models indicate it has anti-addictive and anti-depressive properties. It has also been shown to have neuroprotective qualities, which gives it the potential to treat several neural pathologies, including brain damage induced by oxygen deficiencies (hypoxia).

However, several contradictory studies show that SA can have both anti-depressive and pro-depressive effects, as well as aversive and rewarding properties. In animal models, low doses of SA indicate anti-depressant effects, as well as rewarding effects. In turn, higher doses appear to promote pro-depressive states. These studies suggest that there is very little consistency in the effect of SA, specifically to modulate pro- and anti-depressive states. Researchers suggest such contradictory results may occur as an effect of different dosage concentrations and methods of administration. Nonetheless, these discrepancies need to be taken into account if SA is to be used in a medical setting to treat clinical disorders.

A laboratory trial that tested the effects of SA on cognitive abilities and mood showed no cognitive impairment in any of the ten participants. At a one-month follow-up assessment, participants showed no increase in mood disorders, such as anxiety, depression, fatigue, confusion, or other psychiatric symptoms. However, two months post-treatment, 13% of participants were still experiencing headaches and drastic changes in mood, including uncontrollable laughter or crying (Addy, 2016).

A within-subject, placebo-controlled study on the effects of SA on the brain was conducted using brain imagery techniques to analyze the effects on KORs (Doss et al., 2020). Participants had a long-term history of experience with psychedelics, as well as experience with inhaled substances such as DMT and nitrous oxide. In addition, participants had no history of any psychotic disorders such as schizophrenia and bipolar disorders. The trial included a talk therapy session, where participants were encouraged to discuss the upcoming procedure, as well as expected outcomes. Results showed that SA administered through inhalation showed very similar neural effects as other hallucinogens, such as LSD and psilocybin.

Another study on the potential for SA as a treatment for cocaine addiction was completed and published in 2014 (Kivell et al., 2014). The study focused specifically on *kappa* opioid receptors because of their historical anti-addictive properties. This included SA analogs, such as the traditional KORs, and new analogs. Cocaine is considered a KOR-activating drug, which causes alterations in the *kappa* opioid system. Because these types of drugs reinforce negative feedback loops within the brain, there is a buffer to the increase in dopamine levels in response. This activates the addictive qualities of the drugs, prompting the user to ingest more of the substance to achieve the same results.

Results of the study by Kivell et al. (2014) showed that pure SA was metabolized very quickly, which, as mentioned previously, reduces its suitability for use in clinical trials and within a medical setting. Nonetheless, SA induced a decrease in the effects of cocaine on preclinical models. It also decreased cocaine-seeking behavior in animal models. The duration and dose of SA had differing behavioral effects in animal models. This suggests that acute doses may have aversive effects on reward and motivation by decreasing dopamine release, but these same doses can increase these effects when administered chronically. Low doses of SA have fewer negative effects. This further promotes the anti-addiction effects of SA with fewer side effects than other KOR agonists. Because they retain their anti-addictive properties, KOR agonists that are structurally based on SA have been synthesized as potential alternatives.

Further, Kivell's study (2014) showed that the effects of one of the analogs (MOM Sal B) were similar to that of SA, in that it minimized the search for cocaine-primed rewards but showed no increase in hyperactivity induced by cocaine. In conclusion, it is evident that several SA analogs have the potential to act similarly to SA without negative side effects, maintaining its anti-addictive properties. Additionally, the study confirms the theory that the anti-addiction characteristics of SA occur at concentrations below the threshold, thereby reducing the possibility of negative side effects, such as aversion and sedation.

Maqueda et al. (2015) administered three doses of vaporized SA in increasing doses, as well as a placebo to eight participants, each with previous experience with psychedelics. The study was double-blind and placebo-controlled. Although participants were aware that they were to receive ascending doses, they were not made aware of the placement of the placebo in the trial. Lower doses (0.25 and 0.50 mg) induced somatic sensations; however, higher doses (1 mg) caused severe out-of-body experiences, where participants reported feeling as though they had lost contact with their bodies. Common experiences at lower doses also included pleasant, tingling feelings in the body, a state of peace, and a feeling of personal detachment. Effects were reported a few minutes after administration.

With higher doses, participants reported receiving messages from other entities, an intense tripping experience, a loss of sense of self, and a fading of the physical world. High doses induced effects almost immediately. This study highlighted, once again, the importance of dose sizes, especially if SA is to be considered for use in substance-assisted cognitive therapy. However, it also suggests that administration of SA by vaporization and inhalation is an effective way to control doses and subsequent effects. It also encourages rapid onset psychotropic effects and short duration.

Various computational and biochemical studies have been done on the binding aspects of SA on KORs. These studies hope to clarify the point-mutation alterations and receptor designs that will assist in developing effective analogs with no side effects.

Practical Considerations in Therapy

Currently, SA, its analogs, as well as the plant, are highlighted for clinical use, specifically in pain management and treatment of inflammation. However, there are few studies analyzing the side effects of its use, as well as potential cytotoxicity as a result of misuse or prolonged use. In a review of reports submitted to the California Poison Control System, Coffeen and Pellicer (2019) noted that intentional ingestion of *S. divinorum* resulted in various cardiovascular, gastrointestinal, and neurological effects. However, it must be noted that these reports suggested a concomitant use of the plant with other substances that may have been misused.

Overuse or 'overdose' of SA can lead to a variety of negative responses, including a loss of physical control, and spatiotemporal dislocation (a feeling of being transported to another place or time). These effects can create a frightening experience that may trigger psychotic episodes. Individuals with a history of psychoses may experience prolonged psychotic effects after use.

Also, because of the rapid onset of symptoms that result from inhalation, there are concerns that SA may cause significantly more incapacitating effects, like near-death experiences and amnesia. Other undesirable side effects, which may include aversion, depression, and sedation, have restricted its use in clinical trials and studies. Various studies have suggested that there are serious and negative effects that can occur with the long-term use of SA. These include the following:

- mood swings
- difficulty concentrating
- uncontrollable laughter
- impacts on mental health
- impacts on the ability to make reasonable choices

In another study on rodents, no cardiac, skin, or temperature responses were recorded with acute administration of the substance. Moreover, there were no signs of adverse cardiac effects or significant impacts on any internal organs. Various trials show no significant changes in blood pressure, heart rate, or other physiological activities. These characteristics suggest that SA is safe for use and has a low potential for overdose or cytotoxicity. Further research shows that any histological toxicity is highly time- and dose-dependent, suggesting that SA has a very low toxicology profile.

Much like many other psychedelics, studies suggest that the ability of SA to disrupt neural circuitry can lead to the ability of the user to change their perceptions of previously embedded thought processes.

The few studies that have been done on SA suggest that it is a likely candidate for the treatment of inflammation and pain, and possibly certain psychiatric disorders. However, further research is required to finely identify the actual KOR activity that produces the desired effects of pain relief and anti-inflammatory properties. Analog research shows that it may be possible to synthesize partial KOR agonists, also called functional agonists, that are able to create the anti-addiction effect required without undesirable side effects. The study and development of analogs without unwanted side effects are providing renewed hope that KORs can be produced that have the desired anti-addictive and analgesic properties.

Currently, the opioid crisis that has partly resulted from the misuse of pain medication indicates the great need for an alternative, non-toxic, non-addictive, but effective analgesic. In addition, SA's kappa opioid effects show promise for its use to treat cocaine addiction, as well as depression and other mood disorders. Nonetheless, research does suggest that further in-depth studies are required to better understand the underlying cellular informational pathways used by these compounds before they are used therapeutically.

2C-B AND OTHER PHENETHYLAMINES

Generally, psychedelics can be categorized according to their primary mechanisms in the brain, namely serotonergic or psychoactive hallucinogens, such as LSD and psilocybin, or classic phenethylamines like MDMA and mescaline.

Phenethylamines are naturally occurring organic compounds that have numerous effects on the brain, depending on their chemical structure. They can be found in bacteria, fungi, plants, and animals. Phenethylamines can also be synthesized. Based on slight changes in their chemical structures, they can be stimulants, hallucinogens, or entactogens (chemicals that produce a feeling of connectedness and emotional openness). The family of phenethylamines is known by the general chemical term "2C" drugs. This refers to the ethylene structure of the compounds, which is a two-carbon alkene.

A variety of 2C compounds were synthesized in the 1970s and 1980s by the American chemist and pharmacologist, Alexander Shulgin, primarily for psycho-therapeutic use. However, reports of significant gastrointestinal effects caused its use to be abandoned soon thereafter. The derivative 2C-B first appeared in the mid-1980s as a first-generation synthetic analog of mescaline. By the late 1990s and into the early 2000s, these substances emerged as recreational drugs that were popular in the rave and dance music scene. It was often used as a legal substitute for MDMA. It is commonly referred to as a designer drug, specifically because of its novel and recent synthesis and popularity.

Chemistry and Pharmacology

The phenethylamine class of substances has well-documented psychoactive and stimulant elements. Stimulants include methamphetamines and amphetamines, which are D series. The substitution of the ring series, as described above, results in the creation of what is also known as the 2C series. These phenethylamines are represented by the following key derivatives:

- **2C-B**: 4-bromo-2,5-dimethoxy-β-phenethylamine
- **2C-I**: 4-iodo-2,5-dimethoxy-β-phenethylamine
- **2C-D**: 2,5-dimethoxy-4-methyl-β-phenethylamine
- **2C-E**: 4-ethyl-2,5-dimethoxy-β-phenethylamine
- **2C-T-2**: 4-ethylthio-2,5-dimethoxy-β-phenethylamine
- **2C-T-7**: 2,5-dimethoxy-4-propylthio-β-phenethylamine

Chemically, all phenethylamines are composed of a phenethylamine backbone, which is two phenyl groups that are attached to a carbon and nitrogen atom, which form the ethylamine. Phenethylamine derivatives, also called substituted phenethylamines, are formed by substituting or replacing one or more of the hydrogen atoms in the core structure. It is these simple substitutions that vary the neurological effects of the substances. Several analogs, specifically 2C-B-Fly-NBOMe (2-(8-bromo-2,3,6,7-tetrahydrobenzo [1,2-b:4,5-b']difuran-4-yl)-*N*-[(2-methoxybenzyl]ethan-1-amine), known commonly as NBOMe, can have severe health effects on users, including toxicity and death, even in small amounts (Syrová et al., 2023).

For example, the 2C series is only different from the 2D series by a slight modification of the chemical structure. However, this small change is what causes the hallucinogenic properties of the 2C derivatives. The presence of methoxy groups on the aromatic ring, as well as the separation of the primary amine functionality from the phenyl ring, produces psychoative effects that can range from a low-level stimulant effect at small doses, to a potent hallucinogenic and entactogenic or empathogenic effect at higher doses. By substituting various positions on the benzene ring of a phenethylamine, compounds such as mescaline can be produced, while substituting on the side chains can produce amphetamines. By adding a hydroxyl group, compounds like norepinephrine are produced.

2C substances are known as novel (or new) psychoactive substances (NPS). This is a group of different substances that have the potential to cause significant personal, social, and health impacts, on both individuals that use them, as well as on society in general as a result of this use. Recreational users consider NPS substances as "legal highs." They are either newly developed and synthesized pharmaceutical substances or analogs of existing controlled drugs. They have been created to "legally" mimic the hallucinogenic and stimulant effects and actions of licensed medication and controlled substances.

The 2C-D derivative, which has the lowest potency of the group, is more than eight times more potent than mescaline. The 2C-B derivative is isolated from mescaline and is a hallucinogenic substance that can cause the subjective and emotional effects of classical hallucinogens. It is one of the most commonly used serotonergic psychedelics of the family of phenethylamines. As of 2015, González et al. (2015) reported that 2C-B was one of the most widely used and novel psychoactive, substances available. However, because information on its pharmacology and chemistry, as well as its impacts on the human brain, is limited, concerns have arisen over its use, especially its chronic and acute effects.

In general, most phenethylamines have an affinity for serotonin 5-HT2 receptors, interrupting the uptake of serotonin, dopamine, and noradrenaline. In other words, they modulate monoamine neurotransmitters. 2C derivatives act as agonists or antagonists at these receptor sites. Specifically, the 2C-B analog shows a high affinity to the 5-HT_{2A}/5-HT_{2C} receptors, with a behavioral effect of similar 5-HT_{2A} agonists. In addition, more recent research has suggested that 2C-B also has a secondary selectivity for receptors dopamine, adrenergic, and histamine. 2C-B has similar receptor activity to MDMA and norepinephrine in its ability to inhibit reuptake of the serotonin reuptake transporter.

Stimulant phenethylamines control and regulate neurotransmitters such as dopamine, serotonin, and norepinephrine. They have similar effects to other substances such as cocaine, ecstasy, and methamphetamine. By inhibiting the uptake of dopamine and norepinephrine and increasing their release, phenethylamine stimulants like methamphetamine and amphetamine can cause euphoria, alertness, and arousal.

Hallucinogenic phenethylamines mimic the effects of substances such as DMT and LSD by mediating serotonin-specific receptors. Mescaline is a naturally occurring phenethylamine that causes alterations in perception, as well as feelings of euphoria (see Chapter 8). However, because of their close similarity in structure to stimulant phenethylamines, even hallucinogenic phenethylamines can have a residual stimulant effect.

Empathogenic or entactogenic phenethylamines, like MDMA (see Chapter 4), encourage the release of large amounts of neurotransmitters norepinephrine, dopamine, and serotonin to cause feelings of euphoria, emotional closeness, and a heightening of sensory perceptions, including tactile, auditory, and visual senses.

Therapeutic Applications of 2C-B and Other Phenethylamines

Currently, there are no reported therapeutic applications of 2C-B substances. Information on other phenethylamines, such as MDMA and mescaline can be found in other chapters within the book.

2C-B is most commonly taken as a recreational drug, either orally in tablets or as a powder. The effects of these substances can last between four to eight hours, although this is highly dependent on the dose as well as the user's susceptibility to the chemical compounds.

Because it combines hallucinogenic and stimulant effects, 2C-B produces mild psychedelic feelings, paired with experiences of euphoria and emotional openness. This mix of features also reduces the susceptibility of the user to the negative mood states often experienced with pure hallucinogens. While phenethylamines can have both stimulant and hallucinogenic effects, they can also cause body pains, nausea, vomiting, depression, confusion, and dizziness.

As with other psychedelics, the effects of phenethylamines, and in this case 2C-B, are largely influenced by the environment in which the substance is taken (known as the setting), as well as the mindset of the user (called the set). Common and subjectively positive side-effects and experiences with 2C-B include (*Safer using-2C-B*, n.d., para. 5):

- laughing
- an increased sex drive
- profound religious and spiritual experiences
- visual and auditory hallucinations
- altered perceptions of space and time
- enhanced appreciation of music
- enhanced tactile sensations
- enhanced color perception and visual sensations
- euphoria and lifting of moods

Generally, experienced users suggest that doses between 5-15 mg are considered low. Doses of 15-25 mg are considered moderate and doses between 25 and 40 mg are considered strong. Doses over 40 mg can be considered an extremely high or heavy dose, possibly even an overdose.

Further, anecdotal evidence cautions against using 2C-B in conjunction with other substances such as cocaine, amphetamines, or cannabis, as this polydrug use can result in increased anxiety and seizures.

Clinical Studies and Research

In a placebo-controlled, double-blind, within-subject clinical trial, the effects of 2C-B were compared with psilocybin treatments and a placebo control (Mallaroni et al., 2023). Subjective emotional responses, as well as cognitive and cardiovascular effects, were monitored in a sample of 22 healthy participants who had prior experience with psychedelics. Results showed that 2C-B had similar hallucinogenic effects to psilocybin, which diminished after six hours after exposure. This supports the categorization of the substance as a psychedelic drug with entactogenic properties. The study also showed that a higher dose of 2C-B (20 mg) had a lower subjective impairment effect on participants than a slightly lower dose of psilocybin (15 mg) while exerting similar hallucinogenic (neurological) effects. This promotes it as a well-tolerated and safer alternative to psilocybin for use within a medical setting. The study showed that none of the participants experienced any adverse effects from the substance, and it was well-tolerated overall.

In general, while participants experienced similar alterations to visual and bodily perception as that caused by psilocybin, 2C-B exerted less of an effect on waking consciousness experiences, including spiritual events, disembodiment, and auditory alterations. In addition, participants exposed to psilocybin reported significantly higher levels of physical impairment, including impaired motor and cognitive control, as well as a decrease in vigilance. Overall, 2C-B seems to be less threatening to the ego, as well as exerting a less negative effect on the emotional head-space of the participants. However, it provides the user with a greater ability to experience visual and tactile effects. While both 2C-B and psilocybin were able to provide participants with higher levels of euphoria and friendliness than the placebo, psilocybin caused sustained mood imbalances, a greater range in emotional effectiveness, and an increase in incidents of depression. Rather, 2C-B elicited similar positive experiences to MDMA, including feelings of euphoria, vigor, and elation. Experienced users likened the experience to that of dosing a combination of LSD and MDMA.

Participants exhibited similar elevations in blood pressure with both psilocybin and 2C-B, in comparison with the placebo; however, these elevations were not significant or threatening in nature and were considerably lower than those attributed in other studies to MDMA and other amphetamines. In addition, 2C-B did not show any significant differences in heart rate compared to the placebo.

The results of this research are important to determine the safety of use and dose strength when considering 2C-B as an alternative to other psychedelics in substance-assisted therapy for psychiatric disorders. This is especially significant in light of the fact that very few clinical studies have been done on these phenethylamines and their subjective emotional and cognitive, as well as physiological effects.

In a study conducted by González (2015), twenty healthy participants were subjected to a self-administered dose of 2C-B and evaluated for emotional and cardiovascular effects. Results showed that, while participants experienced a significant decrease in feelings of anger, they also were less able to recognize expressions of happiness and developed an increase in reactivity to negative emotions. However, all participants reported feelings of euphoria and well-being, as well as slight hallucinogenic effects.

Following González's 2015 study, his team once again studied the acute pharmacological effects of 2C-B on humans, namely the effects of the drug once it binds to its affinitive receptor (Papaseit Fontanet et al., 2018). An observational analysis was done on 16 healthy participants with a history of and experience with other, similar psychedelic substances. Once again, participants self-administered an oral dose of 2C-B of varying concentrations of 10, 15, or 20 mg. Participants were not aware of the dose they were administering. From there, various cardiovascular effects were noted over a period of six hours, while participants were asked to complete several subjective emotional and cognitive response surveys. Oral samples were taken to measure the concentration of the stress hormone, cortisol, in saliva. Results showed that these acute doses prompted an increase in both blood pressure and heart rate. All participants reported feelings of euphoria and changes in perceptions of their bodies and their surroundings to various degrees, based on the doses. Just under a third of the participants experienced slight hallucinatory effects. The study concluded that 2C-B produces similar psycho-stimulant effects to drugs that also have an affinity for serotonin receptors.

Also in 2018, Nugteren–Van Lonkhuyzen investigated the clinical toxicology of 2C-B. The study aimed to review the severity of toxicity when patients were exposed to low, moderate, and high doses of the substance. The data was taken from a collection of information obtained from a national poison information center; therefore, most information was considered mostly anecdotal and largely based on the patients' accounts and pharmacodynamics of events. The team categorized doses up to 20 mg as low to moderate, and doses above 20 mg (or unknown) as high. A total of 59 patients were sampled.

Patients reported experiencing neurological, mood, and cognitive effects such as aggression, agitation, confusion, anxiety, and hallucinations, as well as physical symptoms of pupil dilation, an increase in heart rate, and high blood pressure. Results of analyses of the reports suggest that users only experienced moderate levels of toxicity, even when dosing at extremely high levels (one patient reported a dose of 192 mg). In general, patients reported experiencing mild cognitive and somatic effects that lasted no more than 24 hours. No deaths were reported over a period of two years of consolidated data. This review confirmed previous studies that suggest 2C-B-related toxicity may be influenced more by differences in the pharmacokinetics (how the substance is metabolized by the body) and pharmocodynamics (how the body reacts biologically to a substance after exposure) of the substance, rather than dose strength.

Practical Considerations in Therapy

Most phenethylamine compounds are categorized as Schedule 1 substances, which means they do not have any medical relevance and provide a potential for misuse. Specifically, 2C-B was classified as a Schedule 1 substance under the Controlled Substances Act as early as 1995, while others, such as 2C-I were only placed on the list in 2012. Currently, the use of 2C-B is controlled in many countries across the world. However, many of the D-series and other 2C-series substances are not under any international control.

Several observational studies have indicated that 2C-B substances are relatively safe for use, with users showing reduced feelings of anger and an increase in speech emotionality (the ability to convey emotional cues through speech), similar to other psychedelics. However, research does suggest that psychoactive effects are highly dose-dependent.

At oral doses of 4 mg, users usually begin to experience threshold effects, where they start to experience feelings of passivity and relaxation, as well as euphoria and an increase in receptivity to aural, olfactory, visual, and tactile sensations. At oral doses of 8-10 mg, users experience full hallucinogenic states, as well as slight stimulant effects. At doses of between 20-40 mg, users report hallucinogenic states similar to that experienced with LSD. However, at doses over 50 mg, users start experiencing fearful and negative hallucinations. The onset of effects occurs between 20 and 30 minutes after exposure, peaking at 1.5 to 2 hours, and lasting up to 8 hours.

Acute doses can cause common side effects such as dehydration, and an increased heart rate, blood pressure, and body temperature. However, chronic use has been known to lead to dependence, misuse, persistent psychosis, and hallucinogen-persisting perception disorder (a clinical condition where exposure to a hallucinogenic substance can cause the user to experience similar perceptual disturbances for months or years following that). Long-term use, specifically of the stimulating phenethylamines, can cause cognitive deficits and neurotoxicity.

As it stands, there are no confirmed deaths due to exposure to 2C-B. However, there have been reports of fatalities due to cardiac arrest after ingestion of phenethylamines, although the chemical was taken in conjunction with other substances (polydrug use). Nugteren–Van Lonkhuyzen et al. (2018) note that there are only two documented cases of complete intoxication with 2C-B in the US:

- A 19-year-old ingested an extremely high dose of the substance and experienced multiple generalized tonic-clonic seizures. They were successfully discharged 24 hours later with no further impacts.
- An 18-year-old presented with serotonin syndrome, tonic seizures, and severe brain edema, which caused persistent neurologic impairment.

Despite these reports, there are no authorized studies that have investigated the effect of doses of 2C-B over 20 mg. Because of this, the clinical use of the substance is uncertain, as well as its safety. This has hampered the authorization of the drug for use in medical settings, as well as decision-making regarding its use.

Research notes that D-series phenethylamines may have a longer-lasting and more potent effect than other members of the family. Adverse effects of over-exposure to D-series phenethylamines include seizures, liver and kidney failure, agitation, tachycardia, hallucinations, and restriction of blood flow in the limbs.

Anecdotal reports suggest that over-exposure to 2C-B can cause heart palpitations, weakness, agitation, and hypertension; however, in these cases, it is possible the substance was contaminated with other toxic synthetics. Nugteren–Van Lonkhuyzen (2018) lists the following psychotropic and sympathomimetic (physiological effects on the sympathetic nervous system) effects at high doses of the substance:

- marked hallucinations
- increased blood pressure
- increased heart rate
- overheating
- visual disturbances
- trembling
- sweating
- abdominal pain
- jaw clenching
- nausea

While severe and toxic long-term effects are not commonly reported, there is anecdotal evidence that severe effects can occur with high doses of 2C-B, including long-term psychosis, sleeplessness as a result of anxiety, diseases of the blood vessels, and long-term disturbances in visual perception.

To date, there have been no complete analyses and characterizations of the effects of 2C-B, especially in controlled clinical trials conducted by medical professionals. However, several studies have indicated that 2C-B causes minor effects similar to other psychedelics and stimulants. In general, researchers suggest treating and managing the effects of 2C-B as they would ecstasy or MDMA.

Further pharmacokinetic trials need to be conducted to determine dose effectiveness and safety levels of 2C-B if it is to be considered for use in a medical setting. As it stands, there are no reported cases of clinical trials that use 2C-B to treat any medical conditions. While there is an increased interest in the use of phenethylamines, specifically 2C-B and other mixed empathogenic-psychedelic substances, in the treatment of various psychiatric disorders, scientific evidence is lacking.

However, 2C-B can confidently be considered a safer substitute for other psychedelics, such as psilocybin and LSD for the treatment of various psychiatric disorders. This is specifically because the few clinical studies that have been done show that 2C-B has fewer negative psychiatric and physiological side effects when administered, even at high doses.

COMPARATIVE ANALYSIS

For centuries, and even millennia, various cultures around the world have harnessed the inherent psychedelic qualities of plants, fungi, and animals to enhance spiritual and religious rituals and ceremonies. These naturally occurring compounds have also been used to treat other medical conditions, diseases, and illnesses. Natural psychedelics have played (and still do play) a key role in connecting shamans and other practitioners to the spiritual world, gods, and other entities, as well as to their inner selves. To date, many traditions incorporate natural psychedelics into religious and spiritual practices.

In recent years, these psychedelics started being used recreationally as mood enhancements and as hallucinogens. As a result of this use and potential misuse, many of them have been placed under substance control in various countries, the US included. Between the 1950s and 1970s, psychedelics were liberally used in recreational settings; however, they were also being used in a medical setting to treat alcohol dependency, anxiety, neuroses, and other forms of psychiatric disorders. During that time, researchers were insistent that the therapeutic effects of psychedelics were primarily a result of the set (mindset) and setting (surroundings and context) of the user. However, new technologies developed in the 1980s and 1990s, specifically brain imagery of various sorts, have allowed researchers to analyze the mechanisms of psychedelics in the brain. Neurologists are now able to pinpoint the point of action of the psychedelics within the neurons and other parts of the central nervous system to better understand the effect they have, both psychologically and physiologically.

In addition to recreational use, researchers have started recognizing the therapeutic benefit of psychedelics, especially in conjunction with psychotherapy (known as psychedelic-assisted therapies, or PAT), to treat numerous psychiatric disorders, such as depression, anxiety, PTSD, OCD, and others. In specific, classic psychedelics, such as LSD and psilocybin are now known to enhance neuroplasticity, allowing those with psychiatric disorders to disrupt negative thought patterns and develop different neural circuitry. This has further led to the research into and development of synthetic analogs that can be manipulated to maximize the therapeutic benefits of the substances but reduce potential negative side effects.

While clinical studies have been conducted on the treatment of psychiatric and other conditions with psychedelics, many of them are still classified as Schedule 1 and 2 substances, which severely limits their use in medical settings. This is partly because of the potential for abuse and the lack of pharmacological standards for administration. Nonetheless, the positive results achieved in many of these studies, as well as long-term benefits, have prompted a surge in these trials. Clinical trials have clearly demonstrated that psychedelics are able to enhance neural sensitivity, flexibility, and suggestability, allowing users to accept and process bottom-up information. In other words, PAT allows patients to modify previously rigid behavior, thought patterns, and emotional reactions.

Nonetheless, Cavarra (2022) stresses the importance of developing a set of guidelines, standards, and regulations that will allow PAT to be implemented effectively, which has not yet been achieved. Additionally, the role of trained, medical professionals, specifically therapists, is not clearly defined in PAT.

A review of PAT was done by Cavarra et al. (2022), comparing the clinical studies conducted, and how they are practiced. The review is summarized in the table below.

Type of PAT and number of papers reviewed	Psychedelic substance	Process	Disorder/illness
KEP (5)	Ketamine (NMDA receptor antagonist)	Three-stage treatment: 1. Preparation (three months), whereby the participants discuss goals and motives for the treatment. Participants establish positive expectations. 2. The drug session (once-off). Ketamine is administered to achieve a psychedelic experience of up to one hour. Soothing instrumental music is played. A trained therapist engages with the patient throughout the session, encouraging a positive experience. 3. Group psychotherapy sessions (the day after the treatment). Participants are encouraged to integrate their experience, interpret it, and discuss lifestyle changes to consider going forward.	• End-of-life distress • Heroin dependency • AUD
Medication-Assisted Psychotherapy (5)	• LSD(5-HT2a serotonin receptor agonist) • Dipropyltryptamine (DPT) (5-HT1a serotonin receptor agonist)	4. Over two to three weeks, patients are exposed to 6 to 12 hours of preparatory psychotherapy. The therapy focuses on current goals and issues that need resolving. 5. The treatment is administered in the presence of two trained facilitators (once-off sessions). 6. Patients are encouraged to engage in acts that convey comfort and unity, such as holding hands. Toward the end of the session, patients discuss their experiences. 7. The treatment is followed by three separate integration sessions.	• Palliative care • Depression • Anxiety • Pain • End-of-life distress
MDMA-Assisted Psychotherapy (MDMA-AP) (12)	MDMA (5-HT1a serotonin receptor agonist)	8. Two to three preparation sessions, where participants are encouraged to engage with inner healing to confront past traumas. 9. One to three drug-treatment sessions modulated by two facilitators. Facilitators need to be specifically trained for MDMA-AP. The room in which treatment takes place needs to be private, comfortable, and quiet. Drug treatment sessions take place several weeks apart. Sessions last for between six and eight hours. 10. Two to three integration sessions per treatment session. These sessions last for between 60 and 90 minutes and aim to integrate the MDMA-AP experience with daily life.	• PTSD • AUD

Type of PAT and number of papers reviewed	Psychedelic substance	Process	Disorder/illness
Preparation, support, integration (PSI) (7)	Psilocybin (5-HT2a serotonin receptor agonist)	11. One preparation session of four hours. The aim is to build a trusting relationship between the patient and the therapist. Patients are also encouraged to explore their history and possible reasons for their depression. They are informed of the drug treatment process and what to expect from the experience. 12. Two drug-treatment sessions. Patients are given a low dose at the first session, which is monitored by two psychiatrists. The psychiatrists are tasked with ensuring a smooth session and monitoring the patient's physiological and psychological state. The second session takes place one week after the first. 13. One integration session. This session takes place remotely, three months post-treatment. Here, the patient works to integrate their treatment experience with daily life. Other follow-ups may include email assessments.	Treatment Resistant Depression (TRD)
Trauma Interventions using Mindfulness-Based Extinction and Reconsideration (TIMBER) (2)	Ketamine (NMDA receptor antagonist)	14. An hour preparation session made up of a 10-minute period where previous traumatic events are recalled 15. Following that is a 40-minute transfusion period, where ketamine is administered. During the treatment, patients are encouraged to employ mindfulness, breathing, and observation to recall pleasant, calming memories. 16. The session is concluded with a 10-minute mindfulness procedure, whereby the patient is encouraged to lessen any negative emotions that have arisen, such as anger, anxiety, or frustration.	PTSD
Cognitive behavioral therapy (CBT) (7)	• Psilocybin (5-HT2a serotonin receptor agonist) • Ketamine (NMDA receptor antagonist)	17. Four preparation sessions that involve guided imagery and relaxation. 18. Two to three drug-treatment sessions, facilitated by at least one trained therapist. 19. One integration session per drug treatment session.	• Tobacco dependence. • Opioid dependence • TRD
Motivational Enhancement Therapy (MET) **-5**	• Psilocybin (5-HT2a serotonin receptor agonist) • Ketamine (NMDA receptor antagonist)	20. One preparation session before each drug-treatment session for psilocybin therapy. 21. One debriefing session after each drug-treatment session for psilocybin therapy. 22. No formal preparation sessions for ketamine therapy; however, patients are asked to set goals before the treatment.	• AUD

Psilocybin

Various randomized-controlled trials (RCTs) showed that patients with life-threatening cancer, suffering from anxiety and depression as a result, experienced a significant improvement in mood and well-being after a single psilocybin session. These changes, which occurred rapidly after the session, were further maintained more than six months after treatment. A follow-up on a sub-sample of the participants more than four years post-treatment indicated that just less than half of the study group still reported a remission in their symptoms.

In a study on the effectiveness of PSI to treat TRD, patients showed improvements in symptoms up to six months after the treatment. There were significant increases in the patients' abilities to be open to new experiences, as well as engage in social settings. Patients also experienced a significant decrease in emotional instability and susceptibility to engage with negative thought patterns.

A growing set of research has confirmed that, overall, psilocybin is safe for use and well-tolerated in a therapeutic setting. A study by Studeras (2011) looked at acute, short-term, and long-term effects of psilocybin on healthy participants. In a sample of 110 participants, a significant proportion of the group subjectively reported the experience as "pleasurable," "enriching," and "non-threatening" (Studeras, 2011, p. 1). Only a small proportion of the group experienced any adverse effects (anxiety and panic), although this was only at the highest doses. In addition, these negative effects were effectively managed through interpersonal support networks and psychotherapy. Participants showed no long-term side effects of the substance, which could include substance abuse, prolonged psychosis, persisting perception disorders, or other cognitive impairments. This study concluded that, provided psilocybin is administered in moderate doses in a safe, controlled, and therapeutic environment to healthy, well-adjusted participants, there should be no adverse side effects.

MDMA

Several clinical trials have analyzed the effectiveness of MDMA-AP to treat PTSD. In an RCT conducted with patients suffering from PTSD, results showed that 72% of the participants showed a decrease in symptoms and were twice as likely to experience a remission of symptoms for as long as one to six years post-treatment. Another study looked at the use of MDMA-AP to treat patients with AUD. A sample of 14 patients showed a significant decrease in the volume of alcohol consumed per week, with more than 60% showing abstinence from alcohol nine months post-treatment (Cavarra et al., 2022). Further, MDMA-AP was used to potentially treat patients suffering from end-of-life distress or anxiety. While the results were not significant, all 18 patients showed a decrease in symptoms of anxiety.

Currently, MDMA is the closest psychedelic to receiving FDA approval for use in PAT. The results of a large, three-phase clinical trial on patients suffering from severe PTSD as a result of trauma, combat, and abuse, show that none of the participants experienced any serious adverse effects from the treatment. Rather, more than 65% of the participants in the treatment group reported a complete remission of PTSD symptoms, while 88% of them showed a significant reduction in symptoms (Mitchell, et al. 2021). The study successfully concluded that MDMA-assisted therapy can potentially provide a breakthrough treatment for PTSD and warrants further research.

Because MDMA is an entactogen (also called an empathogen) it elicits feelings of empathy and closeness for oneself and others. This is mainly a result of the release of feel-good hormones, such as oxytocin, as well as serotonin and dopamine. These neurotransmitters are responsible for improving mood and increasing sociability. In addition to this, the administration of MDMA decreases activity in the amygdala. This, in turn, allows patients engaging in therapy to do so without any fear or anxiety of past traumatic memories emerging and allows them to deal with challenges and fears openly and calmly.

Ketamine

In an RCT study conducted with 10 patients suffering from PTSD, the TIMBER process was employed to assist in reducing symptoms of the disorder. Results showed that patients who received the ketamine-assisted therapy had improvements in symptoms such as depression and anxiety, although these were not significant.

Another study on the use of KEP to treat patients suffering from AUD showed that almost 70% of the 86 participants were in remission from their symptoms up to one year post-treatment (Cavarra et al., 2022). A similar study conducted on 111 patients with AUD showed that more than 65% of participants from the treatment group were abstinent from alcohol a year after the treatment session; this is in comparison with only 24% of participants from the control group (Cavarra et al., 2022).

KEP has also shown to be successful in treating patients with heroin addiction, whereby 85% of a sample of 70 detoxified participants remained abstinent one month post-treatment. At two years following the treatment session, 24% of the group were still free from the drug (Cavarra et al., 2022). This was significantly higher than those that received the placebo.

Cognitive Behavioral Therapy (CBT) in association with ketamine has proven to be successful in aiding opioid tapering.

Preparation, Drug Sessions, and Integration

Most forms of PAT employ a three-phase approach, namely a preparation stage, a drug-treatment stage, and a final integration stage. These are summarized as follows:

1. **The preparation session/s**: This varies from between two to ten hours, depending on the form of therapy. In this session, the

patient is gently encouraged to develop a rapport with the medical professionals who will guide them through the drug sessions. It is also a chance for the patient to identify prior trauma and past history that they would like addressed by the treatment. The therapeutic mechanisms of the treatment are often discussed so that the patient is aware of the process and expected experiences. Certain therapies explore the motives that led the patient to engage in the treatment. This stage highlights the importance of "set" in psychedelics, namely the mindset of the patient or user and how that impacts the experience; it is a chance for the patient to set intentions for the treatment.

2. The drug session/s: The duration of these sessions varies greatly depending on the psychedelic employed. For example, a ketamine session can only last approximately 45 minutes, while a psilocybin session can last for up to eight hours. Regardless, these sessions are closely facilitated by at least two trained medical professionals, such as psychiatrists or therapists. Rooms are comfortably furnished, and calm music is often played to relax the patients and create a peaceful environment. Once the treatment is administered and before the effects set in, patients are encouraged to keep their intentions in mind and to employ introspection during the session. In certain therapies, where acceptable, facilitators use comforting behaviors such as hand-holding, touching, rocking, or hugging, as well as regulated breathing to guide and comfort patients through and during the session. Patients are encouraged to direct their thoughts toward relevant issues. If needed, facilitators assist the patients in remaining interpretive of their experience and guide them in integrating any spiritual or religious experiences into their session.

3. The integration session/s: This final but important stage gives the patient an opportunity to integrate their experiences during treatment into their daily lives and apply this insight to deal with their traumas or disorders. Patients are encouraged to consolidate their experiences by journaling or talking through them with a therapist. In most cases, the integration sessions are facilitated by the trained professionals who were involved in the entire PAT.

> These professionals are often existentially-oriented psychotherapists who are trained in psychedelic experiences that generate meaning and spiritual significance. In certain situations, an ongoing process of integration and therapy is encouraged.

While preparation plays an integral role in many classical PATs, research has highlighted the significance of "set" and how this may influence outcomes. Experts suggest that placing too much emphasis on intention setting in the preparation phase may outweigh the actual experience and insights gained during the process. This is especially important to consider, what with the extent to which expectations dictate the effectiveness of interventions. In addition, by carefully curating experiences with structured models, experts are concerned that these strict processes may disrupt the natural progress of the experience. Because of the high level of suggestibility that is produced when using psychedelics, patients are vulnerable to external influence, whether it be from the environment or facilitators. That is why it is crucial that facilitators and therapists be highly aware of their behavior and actions during the sessions.

For decades, the scientific, medical, and clinical interest in the use of psychedelics to treat various psychiatric and other disorders and illnesses, has waned, mainly due to their categorization as illegal and controlled substances. However, in recent years, interest in their use has grown, especially as the incidences of psychiatric disorders such as depression and anxiety soar worldwide and many sufferers often no longer find relief from conventional treatments. Pharmaceuticals such as serotonin reuptake inhibitors (SSRIs) and serotonin-norepinephrine reuptake inhibitors (SNRIs) have traditionally been used to treat depression, anxiety, and PTSD, among other mental health conditions. However, these drugs often take weeks or months to show effect and have various serious and sometimes harmful side effects.

In comparison, psychedelics often provide immediate relief from symptoms after a single dose. This is especially significant for sufferers of suicidal ideation who are in dire need of rapid results from treatment. The reported success of psychedelics in decreasing and (sometimes) eliminating symptoms of psychiatric disorders is mostly attributed to their actions on the NMDA receptors of the glutamate system. SSRIs and SNRIs act as agonists, inhibiting the uptake of neurotransmitters like serotonin into the presynaptic axon terminal. This prevents serotonin from being reabsorbed for recycling and degradation and leaves more of the neurotransmitter available to act on receptors in the postsynaptic neuron.

Certain psychedelics have similar serotonin agonist characteristics; however, they also act indirectly on the glutamate system. By inhibiting serotonin from entering the presynaptic axon and allowing it to remain between that and the postsynaptic neuron, the release of glutamate is encouraged. An increase in glutamate activates the NMDA receptors, thereby promoting synaptic plasticity and alterations in the neural network. These mechanisms are what provide the common anti-anxiety and anti-depressive properties of psychedelics. Additionally, psychedelics tend to allow patients to address deep, underlying trauma and experiences, providing them with the understanding required to process and release those issues. This is especially pertinent for patients who find no relief from traditional medications.

The appeal of psychedelic-assisted therapy is in its multi-faceted and holistic approach to treatment. While traditional treatment relies primarily on medications, with therapy playing a minor and often missing role, PAT focuses largely on therapy as the driving feature behind the treatment. By combining both drug treatment and therapy, the PAT process enhances healing and recovery and encourages personal growth and emotional and spiritual integration into daily life.

Overall, the effectiveness of psychedelics is attributed not only to their biochemical effect on the brain but on the integrated approach that encourages patients to shift their perspectives to more positive, beneficial viewpoints.

Luoma. (2020) reviewed nine randomized, placebo-controlled clinical studies that compared the effectiveness of various psychedelics on the treatment of several psychiatric disorders. Substances used included LSD, ayahuasca, and MDMA. Results showed that all three psychedelics were effective in minimizing symptoms of PTSD, anxiety, and depression associated with life-threatening illness, major depressive disorder, and social anxiety, in comparison with the placebo control. This review encapsulates the resounding perspective of PATs: Provided psychedelics are used responsibly in controlled, facilitated environments and sessions, they can be invaluable in treating and assisting patients who suffer from numerous psychiatric and physiological disorders and illnesses.

Case Studies

Ben—a 48-Year-Old Black Police Officer With PTSD

Ben* was a police officer based at the New York Police Department. For almost two decades, this husband and father of two had been working on patrol, responding to domestic abuse reports, shootings, burglaries, and public misconduct. One morning, he responded to a call for a possible explosive device in a building on one of the city's main streets. Ben attended to the call, and on arrival at the scene, spoke to a witness who described seeing a suspicious individual enter the building with what appeared to be a bomb. Ben immediately notified the relevant authorities and took it upon himself to empty the building of civilians. After about ten minutes, Ben was leaving the entrance to the building with a family of four that resided there. Within seconds of leaving the building, the explosive device detonated, causing a moderately destructive explosion. Ben and the family of four were thrown several feet from the building. Unfortunately, the blast had injured a number of the building's residents, as well as fatally harming a passing pedestrian. Ben suffered a serious traumatic brain injury.

For months following the event, Ben suffered from episodes of depression, interspersed with periods of extreme anxiety. He started struggling with his day-to-day activities. Seemingly simple tasks, such as cooking dinner, or driving to the store, filled him with so much anxiety that he would be unable to complete the chore. Minor annoyances sent him into fits of rage, and he suffered from overwhelming insomnia. Ben's quality of life was rapidly deteriorating. He was slowly distancing himself from his wife and children and was starting to perform badly at work. Loud noises caused Ben to startle excessively, to the extent that a back-firing car caused a severe panic attack that landed him in the hospital.

Ben was subsequently diagnosed with PTSD. He was placed on medication and told to attend several therapy sessions with a local psychologist. Ben diligently took his medication and attended months of psychotherapy. After six months, Ben could no longer afford the therapy but continued with his medication. However, a year later, Ben was still suffering from insomnia, panic attacks, severe anxiety, and mild depression. He spent the next two years going between medications in a desperate attempt to solve his illness. Nothing seemed to work. Finally, a colleague told Ben about a new treatment for PTSD that involved MDMA, (MAP). MDMA, a psychedelic that acts to inhibit the reuptake of serotonin would be used in conjunction with psychotherapy to address Ben's symptoms of PTSD. After a few months of preparation, and a once-off dose of the psychedelic, Ben attended a few group therapy sessions. Six months post-treatment, Ben's PTSD symptoms had all but disappeared. He continued to engage with the therapist who had facilitated his treatment and was well on his way to recovery.

Anna—a 35-Year-Old Latin American Housewife With Depression

Anna* was a middle-aged housewife with four sons and a disabled husband. She and her family lived on a middle-class city block in the heart of Seattle. Five years prior, Anna's husband had been permanently disabled during a workplace accident. He was no longer able to work and was receiving a state disability pension. However, the family often battled to get through the month with a limited pension. Oftentimes, Anna was forced to choose between essentials such as heating for their small apartment or food for her family. For the past three years, Anna had been struggling with feelings of sadness, hopelessness, and worthlessness. Coupled with this was a loss of energy, insomnia, and a decreased interest in activities that Anna had previously enjoyed. She spent much of her time in a depressed mood, unable to complete simple tasks such as making the bed or doing the dishes. Frighteningly, Anna started having thoughts of ending her own life. One day, when opening up to her husband about these difficult emotions she was experiencing, he convinced her to see a doctor.

Desperate for help, Anna agreed. She made an appointment with her local general practitioner, who sent her directly to his colleague, a respected and well-known psychiatrist. The psychiatrist diagnosed Anna with major depressive disorder (MDD). After extensive discussions, Anna and her psychiatrist decided to enroll her in a unique and novel clinical trial that showed promising results for those suffering from MDD and treatment-resistant depression. The trial involved the combination of a psychedelic substance called psilocybin and talk therapy. Anna would be required to attend two supervised psilocybin sessions, each followed by intensive and integrative psychotherapy sessions. These sessions would be preempted by a four-hour-long preparation and expectation session, where Anna would work closely with a trained facilitator to discuss the expected outcomes of the treatment.

Just one week after the first session, Anna reported obvious reductions in her symptoms. The periods of depression were shortening, and Anna was starting to feel interest and enjoyment in life once again. These improvements continued after the second session and lasted for several months post-treatment. Anna continued to work with a psychologist, attending monthly sessions where she worked through the issues that had brought on her depression. The psilocybin treatment allowed Anna to fearlessly acknowledge and address her challenges and issues. Talk therapy assisted Anna in dealing with those issues, prompting her to heal. Two years later, Anna's symp toms were in remission, and she was able to successfully stave off depressive moods.

James—a 28-Year-Old Caucasian Veteran With PTSD and AUD

James had been serving in the military for the past eight years. Originally from the Midwest, James had been deployed to several combat zones while enlisted. James' fellow soldiers knew him as a dedicated, disciplined person, with a strong sense of loyalty and ethics. James enjoyed physical exercise and spent every morning completing a strenuous fitness program he had designed. When first deployed, James landed in an active war zone. Within a few days, he experienced several traumatic events. One of his closest friends was fatally wounded by an explosive device and he was constantly exposed to intense combat situations. Nonetheless, James worked to compartmentalize his emotions so that he could successfully continue with his deployment. He noticed that his experiences had a profound effect on him but accepted this as a part of life in the military.

His second deployment was relatively less traumatic, and James made it through with no serious injuries or losses to his company. However, on his third deployment, James was involved in a particularly horrific motor vehicle accident. En route to a patrol base, James and some of his close friends encountered a roadside bomb. The bomb exploded, flipping the vehicle they were in and killing three of his friends. James suffered a broken leg and several serious lacerations. He was otherwise unharmed. However, several days after the incident, James started experiencing vivid, recurring nightmares. These nightmares left him afraid to sleep, which worsened into a case of severe insomnia. Despite being in a state of constant fatigue due to lack of sleep, James found himself hypervigilant toward his environment. This led to extreme anxiety and regular panic attacks.

Shortly after returning home, James found it difficult to re-integrate. Determined that this was the natural consequence of deployment, James did not see fit to seek therapy. Doing his best to push through the constant panic attacks, lack of sleep, and flashbacks of his experiences in the combat zone, James started flagging under the pressure of adjusting to civilian life. A quick Google search suggested that James was suffering from PTSD. Nonetheless, he had heard of many of his comrades who suffered from similar symptoms and had managed to work through them.

Unfortunately, as his symptoms worsened, James turned to alcohol in a last attempt to cope. At first, drinking helped him to calm down and alleviated his severe anxiety. It kept intrusive thoughts and memories away and helped him to relax for the first time in many months. However, what started as social drinking soon turned into a daily habit. His drinking started affecting his relationships with friends and family, and he was fired from his job. It was only after he received a DUI that he realized he needed to make a change.

A fellow soldier encouraged James to try KEP. Ketamine is a psychedelic that works to antagonize NMDA receptors. James participated in a lengthy preparation session where he worked through what to expect from the treatment. Thereafter, he was given a single dose of ketamine by a trained medical professional in a carefully controlled environment. He received this treatment six times over the next three months. Following that, James joined a group therapy session to work through his past trauma and resolve the experiences that led to his PTSD and subsequent AUD. His integration therapy helped him develop healthy coping mechanisms. Six months after his last treatment, James was still sober, had found a new job, and reported significant decreases in his PTSD symptoms. James continued with regular counseling sessions, which he credited with his ability to maintain sobriety.

**denotes a pseudonym*

THE FUTURE OF PSYCHEDELIC MEDICINE

Emerging Trends and Research

Statistics show that over 20% of adults in the US suffer from mental illness disorders, while more than 5% experience severe mental illness (Terhune-Hargrove, 2023). Unfortunately, conventional pharmaceutical treatments seem to be failing many sufferers, especially those with treatment-resistant disorders. Further, there is a concern that there are no evident signs of the increase in capacity for more funding or facilities to meet the growing demand for treatment of and relief from psychiatric disorders and illnesses. In fact, while the amount of budget going towards mental health services is always increasing, these increases are not able to keep up with the need. One of the larger issues with traditional treatment methods, specifically the use of anti-depressant and anti-anxiety medications, is that they are not intended for long-term or chronic use. Additionally, the cost of these medications, as well as the cost of psychotherapy, or talk therapy, is often beyond the financial means of those who need it.

There is an urgent need for an alternative to conventional mental health treatment. The surge in interest in the use of psychedelics, particularly in conjunction with psychotherapy, to treat psychiatric disorders has given many sufferers hope for relief from their symptoms.

Many supporters of this form of treatment, however, are concerned that various preconceived notions surrounding psychedelics are restricting public and federal acceptance of their use in PAT. In the past, most psychedelics, both natural and synthetic, were viewed as harmful and only used in recreational situations. They were believed to have no medical significance and were thought to cause harm and disruption to both individual and societal health. This led to many of them being categorized as illegal and restricted for use under the Controlled Substances Act. The War on Drugs facilitated by the Nixon Administration in the 1970s further created a stigma around these substances. This, as well as other anti-drug campaigns in the 1980s, caused a surge in concern about drug abuse. Toward the end of the 1980s, the percentage of the US public that expressed concern over the use of drugs had skyrocketed from 6% to 64% (Terhune-Hargrove, 2023). While psychedelics had been used in earlier years in the treatment of various psychiatric disorders, the War on Drugs campaign largely put a stop to this, with most viewing psychedelic use as taboo. This further halted any research on the substances.

However, interest in the use of psychedelics in drug-assisted therapy has re-emerged, albeit with certain concerns regarding their safety, as well as the potential for misuse. As discussed, the effectiveness of psychedelics in the treatment of psychiatric disorders is largely based on their subjective effects, which allow patients to engage in deep introspection of their past traumas and issues. This requires intensive psychotherapy provided before, during, and after treatment sessions and involves hours of interaction with trained medical professionals. These PAT sessions have been developed through a combination of traditional and indigenous ceremonies and rituals, non-psychedelic therapeutic approaches, and historical research.

To decrease the need for extensive psychotherapy, research is being done into synthetic analogs that have the same biochemical mechanisms as parent psychedelics but with none of the subjective side effects. This would all but eliminate the need for supportive therapy. However, much work is still to be done, especially in terms of the safety of these analogs. Many experts in the field suggest that it is the ability of psychedelics to enhance neural plasticity, combined with psychotherapy, that actually provides results. The concern is that eliminating therapy from the process will negate its effects.

Brennan and Belser (2022) suggest that PATs that include evidence-based therapies (EBTs are practices that are based on scientific evidence) are more effective than basic support models, which exclude elements of non-psychedelic therapies. They summarize the benefits of EBTs as compared with basic support models as follows:

	Benefits	Limitations
Basic support models (BSMs)	• A more subjective analysis of the experience is encouraged. • Participants are able to make their own informed decisions during sessions. • Trained medical professionals are not required. • Cost savings incurred due to non-medical settings.	• Less potential for efficacy. • Sessions are not governed by guidelines and standards, meaning they are less operationalized. • Staff may be untrained and unable to deal with unexpected situations. • Staff may not fully understand the concept of inner healing. • Staff may not be equipped to deal with any physiological and psychological issues that may arise during sessions.
Models inclusive of evidence-based therapies (EBTs)	• There is greater potential for long-term efficacy. • This model is based on scientific trials and research. • Formal guidelines and frameworks are available to practitioners. • Sessions are professionally conducted by trained personnel. • This model incorporates established behavioral therapies.	• Therapists may be required to learn new theories and practices. • Previous experience and expertise in PAT may not be considered valid for therapists. • Is it not clear if the efficacy of EBTs is translated into effectiveness in PATs. • Therapists and patients may be pressured to conform to established methods and practices.

To further improve the use of EBTs in PATs, Brennan and Belser (2022) have developed an integrated model that aims to highlight the benefits of both BSMs and EBTs and eliminate their limitations. Their approach, called EMBARK, is considered a trans-diagnostic and trans-drug model that supports the therapeutic benefits of traditional PAT while incorporating prior skills and knowledge, as well as using evidence-based theories to provide a holistic practice. Their model embraces six key features that distinguish it from other models. This template is based on the use of psilocybin to treat major depressive disorder:

1. **Existential-spiritual:** The profound existential and spiritual experiences that occur with the use of psychedelics are what often give the practice authenticity. It is the potency of these mystical experiences that has been shown to correlate with the improvements of symptoms such as depression, PTSD, anxiety, AUD, and opioid and other substance dependence (including heroin and tobacco). With the EMBARK model, support for the mystical experience is encouraged during the therapy sessions. This is done by creating an environment that nurtures and supports a spiritual or religious experience. The therapist creates a safe space based on the subjective preferences of each patient. During the actual treatment

session, the therapist uses scientific EBTs to guide and assist the patient through the process, all while encouraging the subjective mystical experience. Therapists ensure that their personal beliefs and biases are not imposed on the patient during the session.

2. Mindfulness: The EMBARK model incorporates the introspective basis of BSMs that encourage self-compassion, mindfulness, and self-regulation, with the EBTs that support psychological flexibility (or being fully conscious in the present moment). Therapists employ scientifically based theories to support the metacognitive shifts that occur during PAT sessions. They use these to guide and assist patients as they engage in the mental freedom that arises with the use of psychedelics. These mindfulness activities allow negative and ruminative thought patterns to be disrupted and give way to positive mindsets.

3. Body awareness: EMBARK therapists are taught to integrate somatic psychotherapy approaches to potential embodied phenomena. They are encouraged to prioritize non-invasive and limit touch-based approaches to dealing with patients during embodied treatment events. Therapists are trained to assist and guide patients back toward body awareness by prompting self-soothing behaviors. Patients are also asked to consent to any touching during the sessions and are given express permission to reject any touching before, during, or after treatment.

4. Affective-cognitive: One of the predominant effects of PATs is the ability of patients to experience extreme shifts in moods and emotions. These can swing from severe anxiety and fear to states of euphoria and elation. Patients also report being able to address trauma and issues with clarity and directness that they normally would not have access to. The non-judgmental and non-defensive manner in which these maladaptive thought processes are addressed is what promotes the efficacy of these treatments. Various EBTs that are not based on psychedelics engage in similar practices that highlight self-acceptance. The EMBARK model incorporates the classical EBT-based theory of approaching negative self-perceptions with a prepared method of dealing with them.

It uses the ability of psychedelics to disrupt negative thought patterns and ideations with the traditional behavioral approaches to cement the patients' positive self-images and cultivate acceptance of themselves.

5. Relational: In PAT sessions, the relationship and rapport that is built between the therapist and patient is important. A therapist needs to be trusted to guide the patient through what can be a difficult process. If handled correctly, a therapist can greatly promote a positive and beneficial experience for the patient. In PAT sessions specifically, the interactions between a patient and therapist can be amplified. This can include incidences of transferences, projections, or re-enactments of previous experiences. However, a poorly handled interaction between the therapist and patient can severely impact the process, leading to harmful experiences. Well-trained, experienced therapists can expertly handle what can be a delicate balance between the support and guidance of a patient during a PAT session.

6. Keeping momentum: Many participants who have engaged in PATs report the long-term benefits of the treatment. Patients report feelings of enduring self-confidence, a sense of motivation, and a commitment to the continuation of therapy post-treatment. The EMBARK model takes advantage of the neural plasticity that is prompted through the use of psychedelics to encourage patients to continue pursuing their post-treatment aims and goals. Therapists are trained to work with and support each patient and their subjective experiences.

As the number of clinical trials investigating PAT steadily increases across the world, these substances that were, for many years, considered dangerous and with no medical value, are becoming widely accepted as having strong therapeutic potential. Within the last several years, various states in the US have started working towards legalizing or at least decriminalizing the use of psychedelics, for both therapeutic and recreational use. For example, the state of Oregon recently had a bill passed that legalized the administration of psilocybin in a therapeutic setting by licensed professionals to people 21 years and older. While the substance is not legal for home or recreational use, it can be used legally under the guidance of trained facilitators. Other states are in the process of legalizing other psychedelics such as LSD and MDMA. In addition to this, other federal organizations, such as the US Food and Drug Administration are considering the legalization of psychedelics for PAT, namely psilocybin to treat major depressive disorder and treatment-resistant disorder, and MDMA for PTSD.

Various clinical trials, conducted on the more classic psychedelics like psilocybin, LSD, and mescaline, show that therapeutic, controlled administration of these substances can actually enhance cognitive health and well-being, at the same time reducing psychological distress and suicidal ideation. In addition, there is very little scientific evidence that this kind of controlled use can lead to substance abuse disorders. In fact, substances such as MDMA are often considered to have anti-addictive properties.

Hendricks (2015) surveyed over 190,000 users of psychedelics in the US, both short-term and long-term. The results of the review showed that lifetime users of classic psychedelics exhibited significantly lower rates of psychological distress, suicidal ideation, and suicide attempts. Again, this review suggests that psychedelics have great potential to assist those suffering from psychological disorders, especially to assist in preventing suicide.

While more research is required to determine safety standards, regulations, and guidelines for administration in a therapeutic setting, the scientific evidence for the safety of these substances in a controlled environment is evident. Overall, psychedelics appear to offer a solution to those suffering from both acute and chronic psychiatric disorders, especially offering relief to those with disorders that resist conventional treatment.

Ethical and Societal Implications

Terhune-Hargrove (2023) suggests that there are four critical industry actions that are required to overcome preconceived perceptions of psychedelics, as well as to encourage PAT going forward:

1. **Education and training**: As noted earlier, the public perception of psychedelics is one of the largest barriers to the acceptance of them for use in PAT. Educating the public on their safe use and benefits, especially for those with serious psychiatric disorders can assist in allaying fears and concerns over potential safety issues and the potential for abuse. Social resistance can be overcome with the right awareness and training, which can encourage acceptance and openness to these unique and advanced therapies.

2. **Safety standards**: While several clinical studies have identified and reported on safe doses of psychedelic substances, particularly identifying doses that encourage adverse side effects, there are still concerns over the safety of these substances. Most importantly, these studies have identified the need for substances to be administered with the close facilitation of trained medical staff and in conjunction with rigorous psychotherapy. Carefully regulated and controlled environments and settings are paramount to the success of these treatments. To further promote PATs, guidelines, protocols, and regulations need to be put in place to safeguard patients and ensure a safe and positive experience. Official regulations will also give PAT the credibility required to garner more acceptance from authorities and the public. This will also aid facilities in obtaining funding and grants for further research.

3. **Specialty sites**: Because psychedelic experiences are largely subjective and based on the mindset and environment in which they are administered, it is of great importance that optimal healing settings are offered. Oftentimes, exposure to psychedelics can be an overwhelming and frightening experience, causing feelings of anxiety, fear, and agitation. To facilitate a positive and beneficial experience, PATs need to be offered from carefully curated environments.

Therapy rooms need to be comfortably furnished, calm, and inviting. Small steps, such as offering serene music and covering mirrors, can greatly enhance the overall experience and the success of the therapy. Facilities that offer PAT need to be aware of the setting in which the therapy is offered; in fact, a good start would be to develop detailed guidelines around ideal environments in which to offer PAT.

4. Access: As with many medical treatments, it is of paramount importance to ensure accessibility for all. The goal is to make PAT open and available to everyone, regardless of background or demographics. One way this can be done is through effective training and education. Another way to positively market PAT is to align it carefully with currently available alternative therapies. Given that psychedelics have been used for centuries, and even millennia, in cultural and traditional spiritual and religious ceremonies, it is important to retain that link to cultural history. In addition, many minority groups tend to seek medical assistance, especially for psychiatric disorders and illnesses, from alternative health providers and not traditional mental healthcare practitioners. To ensure the accessibility of PAT to all minority groups, alternative healthcare providers can be involved in promoting these treatments.

CONCLUSION

As the need for more effective, long-term solutions to mental health illnesses increases around the world, more and more sufferers are looking for alternative solutions. While the use of psychedelics has, for many years, been considered taboo and harmful, they are re-appearing as a safe and promising source of relief for psychiatric disorders such as PTSD, anxiety, and depression.

Over the past decade, the number of clinical trials and studies on the use of psychedelic-assisted therapy has surged. The aim of this research is generally to investigate and understand the impact that psychedelics have on the brain, as well as their effect on various cognitive, mood, and behavioral elements, especially for sufferers of psychiatric disorders. Many studies have looked at the administration of different doses of various substances, on mental health patients as well as healthy participants. Most trials are done as randomized, placebo-controlled, double-blind designs, where the effect of treatments can be compared with placebo controls. Certain studies have looked at the mental and physical health impacts of psychedelics on participants with a long-term or lifetime history of use. Others have investigated the long-term impacts of single-dose treatments on the alleviation of symptoms of alcohol use disorder, PTSD, treatment-resistant depression, major depressive disorder, anxiety, and opioid abuse and misuse. Various reviews have been done on data collected at medical and poison centers to analyze anecdotal evidence of the potential effects of psychedelics.

Most, if not all, of these clinical trials have reported on the positive benefits and impacts of these drug-assisted therapies. Few participants experience adverse side effects; if these do occur, they are often at very high doses. Most negative effects are not significant and are largely related to the expected physical and psychiatric side effects of the substances. These include symptoms such as slight anxiety, nausea, vomiting, and agitation. Overall, participants seem to experience positive benefits of the treatment, including a cessation of symptoms that can last for up to two years. In many instances, with the long-term maintenance of psychotherapy, symptoms for these patients have all but disappeared.

Sufferers of chronic psychiatric disorders such as depression and anxiety often have a diminished concentration of neurotransmitters, such as serotonin, in their brains. These neurotransmitters are responsible for modulating and regulating mood and emotions. When serotonin is maintained at balanced levels within the brain, one feels emotionally stable, happier, calmer, and more focused. However, with many sufferers of mental health disorders, serotonin is reabsorbed by neurons in the brain, after which they are recycled or degraded. This means that there is a reduced concentration of these neurotransmitters available to work on aligned receptors in the brain to induce these feelings of well-being. As a result, sufferers tend to develop negative thought patterns and adverse perceptions of the world around them, as well as themselves.

Due to a mechanism in the brain called the default mode network (DMN), which is a circuit that includes regions of the cortex, angular gyrus, and hippocampus, negative patterns of thinking can be cemented into 'default' neural circuits. Overactive and excessive activity within the DMN is often seen in patients who suffer from depression and anxiety. This increased activity leads to persistent patterns of thinking, specifically self-focused, negative thoughts, and rumination, with worry and over-analyzing included. Impaired or poor regulation of this system can also cause issues with emotional regulation, especially in those with anxiety. This leads to a heightened emotional response to external situations and events and an inability to modulate those reactions. In those with depression, poor regulation of the DMN prevents the sufferer from moving from persistent introspection to focus on the external environment and tasks. This keeps them in a cycle of depressive episodes.

Psychedelics appear to exert two key actions on the brain that assist those with depression and anxiety. Firstly, many psychedelics, especially hallucinogens like mescaline, LSD, peyote, and ayahuasca act primarily as serotonin receptor agonists. This means they inhibit the uptake of serotonin into those neurons that recycle and degrade them. This leaves more serotonin neurotransmitters available in the brain to attach to receptors and enforce their actions to promote positive moods and emotional regulation. Secondly, psychedelics appear to disrupt the DMN, disordering the previously cemented neural networks that cause low emotions and negative thought patterns. Coupled with intensive psychotherapy, patients are able to take advantage of these disruptions to analyze past traumas and personal histories and address the reasons behind their negative thinking. Following this, patients are able to establish more positive thought processes, and with sustained work, can further embed these new neural circuits.

Because of these mechanisms, experts suggest that psychedelics are able to 'change the way a person thinks' and how the brain normally operates (Jackson, 2023). Psychedelics alter old brain patterns and assist sufferers to get out of negative thought loops. Research further shows that these disruptions can last for up to three weeks after a single dose of a psychedelic. This gives the patient a lot of time to develop new patterns of thought. Psychedelics also offer patients the opportunity to engage in psychotherapy with complete openness, without the usual fear and anxiety that many sufferers experience. This allows patients to fully address past traumas and fears to overcome them. While patients with depression and anxiety tend to have brain patterns that work slightly in isolation from one another, psychedelics allow these regions and networks to become integrated, releasing that 'locked-in' characteristic.

However, proponents of psychedelic-assisted therapy (PAT) emphasize that treatment should not rely fully on the work of the substances. In fact, PAT should be considered 50-50; in other words, half the work can be done by the psychedelics but the other half falls to the responsibility of the patient to engage in and maintain psychotherapy. While psychedelics have the pharmacological ability to change the way the brain works, the patients need to do the psychological work to break through those fixed thought patterns. This can be a demanding and difficult task, but the positive results have been proven. Ben Sessa, a British psychiatrist who has been working in the field for the past two decades, suggests that the use of PAT is almost directly contradictive of the conventional way of treating mental health disorders. Traditional psychiatry provides patients with medications to treat the symptoms of the disease but places little emphasis on the psychotherapy needed to address the underlying causes. He refers to this as 'maintenance prescribing' (Jackson, 2023).

As with anything, there are risks involved in the use of psychedelics, especially when used by vulnerable patients or those suffering from psychotic disorders like schizophrenia and bipolar mood disorders. There are few reports of overdose or addiction to psychedelics; however, there is always a risk of addiction to any drug. In those with a history of psychotic episodes, or with family that suffers from psychotic episodes, the use of psychedelics can trigger manic or psychotic episodes. In addition to the potential for these negative psychiatric side effects, psychedelics can also have adverse physiological impacts. For example, substances such as ketamine and psilocybin can significantly increase blood pressure and heart rate, while the stimulation of brain activities can prompt seizures in those susceptible to that.

Because of these risks, it is imperative that the facilitators who conduct these PATs are highly trained and aware of all the risks involved. Trained medical professionals need to be able to recognize the signs of any adverse side effects and should be able to mitigate and deal with these. In addition, facilitators need to be able to advise patients on the safe use of the substances, especially if they suffer from comorbidities or illnesses that may interfere with the treatment.

It is these risks and vulnerabilities to PAT that need to be included in strict policies, procedures, and guidelines. Therapy sessions need to be guided by regulations that ensure complete safety for patients. Facilitators should be adept in both conventional psychotherapy, as well as in holistic therapeutic practices, such as meditation, breathwork, mindfulness, and journaling.

Psychedelics were revered for millennia as a pathway to enhance religious ceremonies and achieve a connection with ancestors, spiritual deities, and other worlds. They were also prolifically used to treat numerous physical ailments. However, centuries later, they were vilified, leading to many of them becoming criminalized. For decades thereafter, psychedelics were considered the domain of those who engage in illicit and illegal activities. They were maligned as a threat to the stability and success of society and believed to be considerably harmful and dangerous. Use was outlawed and campaigns were instituted specifically to prevent their popularity.

However, scientifically-backed success stories that have proven their effectiveness in treating some previously 'untreatable' disorders have started changing that mindset. While psychedelics may not be a miracle 'cure-all' for psychiatric disorders, they most definitely display a great potential to offer relief from these illnesses, as long as they are administered correctly. Neurological technology has proven that the brain is 'plastic', in that it has the inherent ability to change, grow, and adapt. It is this characteristic of the brain that makes psychedelics so effective.

The exploration of psychedelics as a treatment for psychiatric disorders has proven to be a profound shift in mental health care. The once-stigmatized substances have been recognized for their potential to alleviate serious, life-threatening, and treatment-resistant psychiatric conditions. The deep insight and emotional breakthroughs offered by these substances can provide sufferers with a transformative experience that can be used to complement traditional talk therapy. However, the potential for addiction and misuse needs to be balanced with the enthusiasm for the potential benefits offered by psychedelics. Nonetheless, the increasing acceptance of psychedelics for use in medical settings symbolizes the dawning of a new era of hope for psychiatric healing.

Dear reader,

Thank you for reading *The Handbook of Medical Hallucinogens.*

If you enjoyed this book, please leave a review where you bought it. It helps more than most people think.

Don't forget your FREE book!

You will also be among the first to know of FREE review copies, discount offers, bonus content, and more.

Go to:

https://www.SFNonfictionBooks.com

Thanks again for your support.

AUTHOR RECOMMENDATIONS

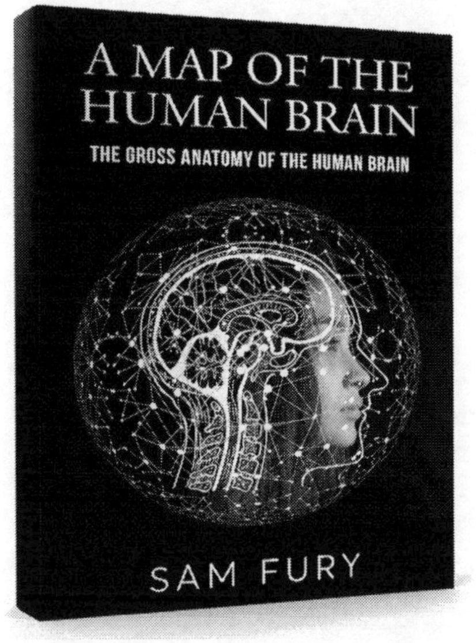

Unlock the Secrets of Your Mind

Nurture your mental well-being, because better brain health begins with knowledge.

Get it now.

www.SFNonfictionBooks.com/a-map-of-the-human-brain

Do You or a Loved One Have ADHD?

Elevate your productivity despite ADHD, because mastering organizational skills can transform your life.

Get it now.

www.SFNonfictionBooks.com/20-productivity-rules-for-adhd-minds

ABOUT SAM FURY

www.SamFury.com

amazon.com/stores/Sam-Fury/author/B00C8Z4U8S
facebook.com/SamFuryOfficial
instagram.com/samfuryofficial
youtube.com/@SamFuryOfficial
x.com/Samfuryoriginal
tiktok.com/@samfuryofficial

REFERENCES

https://classic.clinicaltrials.gov/ct2/show/NCT05523401

https://doi.org/10.1021/acsptsci.1c00018

https://doi.org/10.1016/bs.pbr.2018.08.002

https://doi.org/10.1016/j.cub.2011.10.037

https://doi.org/10.3389/fpsyg.2022.866018

www.awakenyoursoul.co/bwiti-tradition

www.bps.org.uk/psychologist/how-do-hallucinogens-work-brain

https://doi.org/10.1016/j.nrl.2011.07.003

https://doi.org/10.3389/fpsyg.2022.887255

https://doi.org/10.1038/s41591-023-02705-w

https://doi.org/10.2147/JPR.S188619

https://doi.org/10.1038/s41598-023-44193-5

https://doi.org/10.1016/B978-0-12-815607-0.00033-2

www.medicalnewstoday.com/articles/302663#what-is-it

www.medicalnewstoday.com/articles/308850

www.medicalnewstoday.com/articles/306889

www.med.unc.edu/pharm/a-scientific-first-how-psychedelics-bind-to-key-brain-cell-receptor/

https://doi.org/10.2174/1874467211666181010154139

www.fieldtriphealth.com/blog/how-does-ketamine-affect-the-brain

https://doi.org/10.1038/s41598-020-73216-8

https://www.dea.gov/drug-information/drug-scheduling

https://doi.org/10.3389/fpsyg.2023.1131565

https://adf.org.au/drug-facts/empathogens/

www.sciencedirect.com/topics/medicine-and-dentistry/ergot-alka loid

www.forbes.com/sites/milletienne/2023/11/16/the-future-of-psychedelics-explained-by-a-neurologist/?sh=44c75ec88d4a

www.fda.gov/news-events/public-health-focus/expanded-access

www.iceers.org/first-ever-clinical-trial-with-ibogaine-for-opioid-dependency/

https://answers.childrenshospital.org/how-the-antidepressant-keta mine-rapidly-awakens-the-brain-and-why-its-effects-vary-more-in-women/

www.deadiversion.usdoj.gov/drug_chem_info/bromo_dmp.pdf

https://science.howstuffworks.com/magic-mushroom6.htm

https://doi.org/10.1111/j.1360-0443.2006.01511.x

www.qps.com/2022/11/28/from-cultural-rite-to-clinical-trial-exam ining-the-therapeutic-potential-of-mescaline/

www.ucsf.edu/news/2023/05/425246/psychedelic-inspires-new-treatments-addiction-and-depression

https://doi.org/10.1155/2015/643878

https://doi.org/10.1002/jcph.471

https://doi.org/10.1056/NEJMoa2206443

https://my.clevelandclinic.org/health/articles/6734-hallucinogens-lsd-peyote-psilocybin-and-pcp

https://courses.lumenlearning.com/waymaker-psychology/chap ter/reading-hallucinogens/

https://doi.org/10.2174/1570159X16666180125095902

www.news-medical.net/health/Psychedelic-Medicine-Past-Present-and-Future.aspx

https://daily.jstor.org/the-colonization-of-the-ayahuasca-experience/

https://doi.org/10.1177/0269881114565653

Hoffman, A. (1980). LSD—My Problem Child. McGraw-Hill. ISBN 0-07-029325-2

https://doi.org/10.1007/s00213-021-05978-6

https://doi.org/10.1016/j.biopsych.2022.08.025

https://doi.org/10.1002/cpt.2057

www.verywellmind.com/what-is-ecstasy-22091

https://bhoperehab.com/how-do-dissociative-drugs-work/

www.rxlist.com/supplements/iboga.htm

www.sciencedirect.com/topics/medicine-and-dentistry/ibogaine

www.bacp.co.uk/bacp-journals/therapy-today/2023/july-august-2023/the-big-issue/

https://yalebooks.yale.edu/2019/08/06/what-happened-to-mescaline/

www.medicalnewstoday.com/articles/peyote

https://doi.org/10.12688/f1000research.11372.1

www.sciencedirect.com/topics/pharmacology-toxicology-and-pharmaceutical-science/ketamine

https://cdphe.colorado.gov/ketamine-investigatory-review-panel-report-overview

https://greatergood.berkeley.edu/article/item/how_do_psychedelics_work_in_the_brain

https://doi.org/10.1016/B978-0-12-420118-7.00012-3

www.healthline.com/nutrition/ayahuasca#what-it-is

https://edition.cnn.com/2022/06/11/health/psilocybin-brain-changes-life-itself-wellness-scn/index.html

www.healthline.com/health/alpha-brain-waves

https://doi.org/10.1080/02791072.2012.704590

https://doi.org/10.1038/s41386-023-01607-2

https://doi.org/10.1038/npp.2017.86

https://www.history.com/topics/crime/history-of-lsd

https://doi.org/10.1080/02791072.2020.1769878

https://doi.org/10.1080/17425255.2021.1944099

www.sciencedirect.com/topics/chemistry/lysergic-acid-diethy lamide

www.emcdda.europa.eu/publications/drug-profiles/lsd_en

https://doi.org/10.1016/B978-0-12-800212-4.00078-9

https://doi.org/10.3389/fpsyt.2022.1040217

www.technologynetworks.com/neuroscience/articles/lsd-for-anxi ety-a-deep-dive-into-a-new-clinical-trial-366381

https://doi.org/10.1002/cpt.2958

https://doi.org/10.1016/j.xcrm.2022.100906

https://doi.org/10.3389/fphar.2019.00193

https://doi.org/10.1093/in/pyv065

www.chm.bris.ac.uk/motm/lsd/lsd1_text.htm

https://doi.org/10.1016/j.nsa.2024.103938

https://nida.nih.gov/publications/research-reports/mdma-ecstasy-abuse/Introduction

www.emcdda.europa.eu/publications/drug-profiles/mdma_en

https://www.sciencedirect.com/topics/pharmacology-toxicology-and-pharmaceutical-science/mdma

www.drugtopics.com/view/fda-grants-breakthrough-therapy-designation-to-lsd-based-treatment-for-generalized-anxiety-disorder

https://pubchem.ncbi.nlm.nih.gov/compound/4076

www.sciencedirect.com/topics/medicine-and-dentistry/mescaline

https://psychedelics.berkeley.edu/substance/mescaline/

www.bbc.com/travel/article/20200713-basel-the-birthplace-of-hallucinogenic-science

https://doi.org/10.1097/EJA.0000000000000638

https://doi.org/10.1038/s41591-021-01336-3

https://doi.org/10.1038/s41591-023-02565-4

https://recovered.org/hallucinogens/mescaline

www.britannica.com/topic/Native-American-Church

https://doi.org/10.1016/j.annemergmed.2020.04.022.

www.imperial.ac.uk/news/193993/ayahuasca-compound-changes-brainwaves-vivid-waking-dream/

www.verywellmind.com/history-of-psychedelic-use-6745982

https://doi.org/10.1007/s00213-021-05964-y

https://www.ncbi.nlm.nih.gov/pmc/articles/PMC4208627/

https://doi.org/10.3389/fphar.2018.00206

https://doi.org/10.1080/02791072.2015.1128580

https://study.com/academy/lesson/ketamine-pharmacology-chemical-structure.html

www.sciencedirect.com/topics/neuroscience/peyote

www.sciencedirect.com/topics/neuroscience/phenethylamine

www.webmd.com/vitamins/ai/ingredientmono-1274/phenethy lamine-pea

www.unodc.org/LSS/SubstanceGroup/Details/275dd468-75a3-4609-9e96-cc5a2f0da467

https://doi.org/10.1038/s41380-018-0341-9

www.wildflower.org/plants/result.php?id_plant=LOWI

www.harvardmagazine.com/2022/06/vita-richard-evans-schultes

https://doi.org/10.1016/C2013-0-14226-2

www.sciencedirect.com/topics/medicine-and-dentistry/psilocybine

https://nida.nih.gov/research-topics/psychedelic-dissociative-drugs#what-are

www.thelancet.com/journals/eclinm/article/PIIS2589-5370

www.hopkinsmedicine.org/psychiatry/research/psychedelics-research

https://equanimitywellness.co.za/psychedelic-vs-traditional-antide pressants/

www.healthline.com/health/ibogaine-treatment

https://sanescohealth.com/blog/what_is_phenylethylamine/

www.psychiatrictimes.com/view/psychedelics-and-the-future-of-psychiatry

www.cahma.org.au/article/safer-using-2c-b/

www.emcdda.europa.eu/publications/drug-profiles/salvia_en

www.sciencedirect.com/topics/medicine-and-dentistry/salvinorin-a

www.theguardian.com/science/2023/mar/20/psychedelic-brew-ayahuasca-profound-impact-brain-scans-dmt

https://doi.org/10.1001/jamapsychiatry.2017.0080

https://doi.org/10.1177/0269881114552713

https://doi.org/10.5688/aj700598

www.mayoclinic.org/diseases-conditions/serotonin-syndrome/symptoms-causes/syc-20354758

https://doi.org/10.1177/2045125320967197

https://doi.org/10.1001/jamapsychiatry.2022.4101

https://doi.org/10.1016/j.cell.2023.04.010

www.medicalnewstoday.com/articles/mdma-therapy

https://doi.org/10.1177/02698811221092508

Sproul, C. (2021). "Don't kill my buzz, man!"—Explaining the criminalization of psychedelic drugs. Oregon Undergrad Res J, 19(1), 1-53. ISSN: 2160-617X

www.webmd.com/mental-health/addiction/what-is-ayahuasca

www.webmd.com/mental-health/what-is-mdma-assisted-therapy-ptsd

https://doi.org/10.1177/0269881110382466

www.theguardian.com/society/2023/jan/23/ibogaine-iboga-drug-addiction-psychedelic-gabon

https://doi.org/10.3389/fphar.2023.1120419

https://doi.org/10.1038/npp.2017.194

www.appliedclinicaltrialsonline.com/view/breaking-historic-barriers-to-ensure-psychedelic-therapy-s-future

www.webmd.com/mental-health/news/20240326/fda-opens-the-door-clinical-use-lsd

https://doi.org/10.1038/s41598-019-51974-4

https://doi.org/10.1073/pnas.2218949120

www.nycbar.org/videos/the-legal-history-of-psychedelics/

https://doi.org/10.2174/1573399811309999003

www.sciencedirect.com/topics/neuroscience/tryptamine

www.sciencedirect.com/topics/neuroscience/2c-b

https://www.ebi.ac.uk/chebi/searchId.do?chebiId=CHEBI:18397

https://doi.org/10.1007/s00213-018-4988-3

https://doi.org/10.1016/j.neuropharm.2022.109294

https://doi.org/10.31887/DCNS.2001.3.4/fxvollenweider

www.medicalnewstoday.com/articles/ergot-poisoning#what-it-is

https://canyonsantamonica.com/rehab-blog/what-is-the-difference-between-mdma-and-mda/

https://med.stanford.edu/news/all-news/2024/01/ibogaine-ptsd.html

https://doi.org/10.1038/s41598-020-75706-1

www.newscientist.com/definition/ibogaine/

https://physicians.mountsinai.org/news/five-things-to-know-about-mdma-assisted-psychotherapy-for-ptsd

https://addictionresource.com/drugs/salvia/legality/

https://doi.org/10.1186/s12910-016-0087-3

https://doi.org/10.7759/cureus.21944

Made in the USA
Monee, IL
12 November 2026

92b1b1b0-57c8-4793-aac9-91095376bac5R01